Special Presentation Edition

TO

Gerry Evans

with sincere thanks for your contributions to
West Kootenay Power's first

Century of Service

Chairman, President and Chief Executive Officer
West Kootenay Power, 1997

The Presidents

West Kootenay Power
1897 - 1997

1897 - 1899	*Oliver Durant*
1899 - 1903	*Sir Charles Ross*
1903 - 1913	*W. M. Doull*
1913 - 1927	*C. R. Hosmer*
1927 - 1939	*J. J. Warren*
1939 - 1947	*L. A. Campbell*
1947 - 1957	*R. W. Diamond*
1957 - 1965	*R. G. Anderson*
1965 - 1975	*W. F. Gwyer*
1975 - 1978	*P. Hanson*
1978 - 1980	*H. M. Lewis*
1980 - 1993	*J. A. Drennan*
1993 - Present	*D. G. Bacon*

THE BUSINESS OF POWER

HYDRO-ELECTRICITY IN SOUTHEASTERN BRITISH COLUMBIA 1897 – 1997

THE BUSINESS OF POWER

HYDRO-ELECTRICITY IN SOUTHEASTERN BRITISH COLUMBIA 1897 – 1997

JEREMY MOUAT

Sono Nis Press

VICTORIA, BRITISH COLUMBIA, CANADA

Canadian Cataloguing in Publication Data

Mouat, Jeremy, 1950-
The business of power

Includes bibliographical references and index.
ISBN 1-55039-081-3

1. West Kootenay Power and Light Co.—History. 2. Electric utilities—British Columbia—Kootenay Region—History. I. Title.
HD9685.C34W47 1997 333.793'215'097116 C97-910668-0

We acknowledge the support of the Canada Council for the Arts for our publishing program.

Appendices compiled by John Loo.

Published by
SONO NIS PRESS
1725 Blanshard Street
Victoria, BC V8W 2J8 Canada
http://www.islandnet.com/~sononis/

Printed and bound in Canada by
MORRISS PRINTING COMPANY LTD.
Victoria, British Columbia

For H.K.R.,
teacher and friend

An early photograph entitled "Falls of the Kootenai — near Robson, BC."

CONTENTS

PREFACE

> *In no society was electrification a "natural" or a "neutral" process; everywhere it was shaped by complex social, political, technical, and ideological interactions.*
>
> — DAVID E. NYE, *Electrifying America*[1]

In his famous biography, *The Life of Johnson*, James Boswell described a visit to the world's first manufacturer of steam engines. The year was 1776—better known as the date marking the American Revolution—but in a factory just outside of Birmingham, England, Matthew Boulton employed seven hundred people, producing engines that drove a revolution of another sort. The sight impressed Boswell and his account suggests that Boulton too was well aware of the factory's significance. Gesturing to the activity before them, he exclaimed proudly to Boswell, "I sell here, Sir, what all the world desires to have—POWER."[2]

Boulton was right, of course. With the passage of two hundred years we do not so much desire power but depend on it. Energy is central to almost all of our activities, whether they involve work or recreation, creature comforts or daily necessities. The most common form of energy in Canada is electricity. This book examines the history of electricity in one region, southeastern British Columbia, describing the first hundred years of the electrical utility, West Kootenay Power & Light. Like Boulton, the company sells what all the world wants.

Formed in 1897, West Kootenay Power is one of a handful of BC companies to have operated for a century or more. It's worth stressing, however, that had the company not been formed, another company would have emerged to fulfil the same role. Today, electricity would certainly be flowing into the homes and businesses of southeastern British Columbia, however different the circumstances of power development. This raises some interesting questions: why was West Kootenay Power successful, surviving when a handful of competitors failed? Answering this question, we are forced to consider others. For example, did West Kootenay Power build a better plant or use superior technology to harness the river's energy? If so, was this because it had better engineers or managers? Or did it have greater access to more capital? Was its longevity and success simply better timing, the initial advantage of a better economic climate? How useful were its political connections and its links to the Canadian Pacific Railway, the largest and most important business group in Canada? The point of these questions is to suggest that West Kootenay Power's history is not just a narrative of progress with linear growth as the major theme. As David Nye points out in the quoted passage, electrification was a complex process in which many factors came into play.

The following pages hopefully convey something of that complexity. West Kootenay Power's history is not simply a reflection of the personalities of the promoters and entrepreneurs that launched the company; its career bears the mark of unique events, of a specific social and political context. War and depression, good times and bad, all affected the company. On occasion, it could take advantage of unique opportunities; at other times, events outside the region limited its options. And the story involves more than the fortunes of one company. West Kootenay Power's history illustrates in microcosm a fascinating process: the way in which electricity became an integral part of all our lives.

More than twenty years ago I had another brush with electricity. For three years in the mid-1970s, I worked as a skilled labourer for the New Zealand Electricity Department, mostly engaged in road repair and maintenance at a hydro-electric plant. (Occasionally my mother's dire predictions of my likely future as a ditch digger came back to haunt me.) I enjoyed this time, however, for the job allowed me to live in a spectacular location in the northwest corner of the South Island. Only two of us lived up at the dam itself, where we worked a ten-day shift, followed by four days off. My co-worker looked after the dam while I tended the road, although during his days off I carried out the routine chores that were normally his. I also learned something of the dam's history, which was built originally to supply power to a nearby mine, not unlike the initial purpose of West Kootenay Power. My interest in history was just beginning — I took my first university-level history courses by correspondence while working up at the dam — and I never imagined that years later I would again become involved with the world of electricity.

I am grateful to West Kootenay Power for giving me the opportunity to write this book. I commend the company not only for its sense of history but also for its willingness to allow me free and open access to its records. I am particularly grateful to Mike Bradshaw and Donna Bertrand in the company's Trail office for their assistance (and patience). My thanks as well to an important group of historical support workers: Shawn Lamb at the Nelson Museum; Joyce Austin at the Rossland Historical Museum; Brian Young and all the staff at the BC Archives in Victoria; and my colleagues Judy Stady and Sandy Partyka at Athabasca University Library. For help with the photos, I am grateful to Donna, Shawn, Joyce, Jamie Forbes (City Clerk in Trail), and Kelly Nolin at the BC Archives. I also appreciate the assistance of people who kindly allowed me to interview them: my thanks to Josh Smienk for telling me of the Columbia Basin Trust; to Bob Stone, Vice President Finance of Cominco; to Richie Deane, Sr., formerly an electrical engineer with West Kootenay Power and later Chief Electrical Engineer with Cominco; and to Davie McDonald, also a former West Kootenay Power employee. I am particularly grateful to my friend Jack McDonald, whose father's career with West Kootenay Power began last century. Thanks also to Mary Mouat and Don Miller for hospitality over the years, during flying trips to Victoria. Also in Victoria, I am indebted to Ann West of Sono Nis Press for making the process of publication smooth and stress-free (and for only occasionally commenting on my peripatetic lifestyle). Finally, a toast of the best red wine to Lynda Ross, for listening politely to my stories about electricity.

NOTES

[1] David E. Nye, *Electrifying America: Social Meanings of a New Technology, 1880–1940*. Cambridge, MA: MIT Press, 1990, 138-39.

[2] George Birkbeck Hill, ed., *Boswell's Life of Johnson*, Vol. II, *The Life — 1766–1776*, Oxford: Clarendon Press, 1934 (revised and enlarged edition by L. F. Powell), 459. Boswell noted regretfully, "I wish that Johnson had been with us: for it was a scene which I should have been glad to contemplate by his light." For further comment on the visit and the broader context, see H. W. Dickinson, *Matthew Boulton*, Cambridge: Cambridge University Press, 1937, 72-73. As Dickinson notes, Boulton had gone into partnership with James Watt the year before, in 1775, and produced steam engines under Watt's patent.

CHAPTER ONE

The Beginnings of the Electrical Age

When West Kootenay Power's first generator produced electricity in the summer of 1898, hydro-electric enterprises were a relatively new phenomenon. But electricity itself was a common enough concept and by the end of the nineteenth century it had become a familiar feature of North American life. In the late 1890s most cities had some form of electric street lighting, many had electric street railways, and telegraph wires stretched across the continent and beneath the oceans.

The "invention" of electricity can not be identified with any precision nor is there a key date with which we can link its discovery. We do know that gentlemen scientists from the time of England's Elizabeth I studied electricity and the word itself dates from that period. The idea that electrical power could serve as an important source of energy, however, only gained acceptance during the second half of the nineteenth century.

Today, of course, electricity is so widely available and so much a part of our world that we rarely bother to think about it. People use the saying "at the flick of a switch" when they mean that something is easy; it's as simple as turning on electricity. And most of us assume that electricity is there whenever we need it. Power failures — most famously, in November 1965 when New York City as well as much of New England was plunged into darkness — underline our reliance on electricity and remind us of our vulnerability. Yet even power failures have little impact on our assumptions or our behaviour. We may grumble about the inconvenience or make a mental note to buy some candles, matches, and a good flashlight, but hardly anyone would

> ## *Electricity*
>
> An English scientist, William Gilbert, coined the word electricity in a book published in 1600. He wrote in Latin, and so in fact he created a modern Latin word, *electricus*. His choice of word was not accidental: it derived from a Greek word, *elektron*, the word for amber. Like others, Gilbert knew rubbing amber produced static electricity.
>
> — **SIR EDMUND WHITTAKER**
> *A History of the Theories of Aether and Electricity*, Vol. 1, 35

respond to these temporary disruptions of service by getting rid of the numerous electrical appliances that fill the typical North American home.

We take electricity for granted and it is easy to forget that it is a modern phenomenon, one that only recently became a part of daily life. When Benjamin Franklin flew a kite during a thunder storm in 1752, in order to demonstrate the validity of his theory of electricity, few people grasped the concept of electricity let alone its possible practical application. (Franklin immediately established one very useful application, the lightning rod.) Scientists and others gradually moved beyond the initial phase of curious experiment, to speculate on the nature of electricity and its possible applications. Knowledge of electrical phenomena increased dramatically during the nineteenth century, and the closing decades saw a quickening rate of innovation, combined with some shrewd marketing and advertising. That last point is important: one of the things that has changed over the last hundred years is how we think about electricity. In its way, this is as important as the physical presence of power lines along the roadside or the number of electrical outlets in our homes.

Until well into the twentieth century, most Canadian houses did not have access to a constant and reliable source of power. And electricity did not simply appear in our homes like a genie from a bottle: it is a technology with a particular history, a technology that developed slowly, principally over the last 150 years. The growth and success of West Kootenay Power reveals how electricity became common in one region, the southeastern corner of British Columbia. But we should begin at the beginning. To understand why the company's generators began producing electricity when they did, in the summer of 1898, we need to look very briefly at developments in nineteenth century Europe and North America.

The widespread adoption of electricity in Europe and North America is a fascinating story of trial and error, success and failure. This is scarcely surprising, most new technologies experience growing pains, with a particular process gradually eclipsing its competitors. (Think of the battles between audiocassettes and eight-track tapes, between VHS and Beta video-machines, or between MacIntosh and IBM computer systems.) Electricity followed a familiar pattern. The end result is that today most buildings in North America are wired for a particular type of electricity — 110 volts alternating current — and share similar sorts of plugs and light switches. Houses typically have a wide array of electric devices, such as fridges, stereos, television, microwave ovens, and so on. When we visit a retail store to purchase a piece of electrical equipment, we do not have to choose between competing electrical systems. Yet anyone who has travelled overseas will know that the North American system is only one of several possibilities.

The need for electrical standards — from the voltage level to the type of plug — seems obvious enough. Consider the choices confronting people in London, England, for example, prior to the First World War. Ten different electrical frequencies operated in the city in 1913. The world's largest city had twenty-four voltage levels for electrical distribution and more than thirty for transmission, as well as some seventy ways of charging and pricing for power.[1] Understandably, few people welcomed this baffling array of possibilities and there was considerable demand for a uniform system.

Electricity began to be widely discussed in Britain and Europe in the late eighteenth century. At first, this interest was largely driven by curiosity. Experiments were often as much public spectacle as they were serious investigation. But they had a cumulative effect, nonetheless. Even the discovery that the limbs of dead frogs would jerk in reaction to an electric current led to extensive work on the construction

London, 1913

Londoners who could afford electricity toasted bread in the morning with one kind, lit their offices with another, visited associates in nearby office buildings using still another variety, and walked home along streets that were illuminated by yet another kind. This was a nightmare for the supplier of electric lamps and appliances; interesting problems arose when one moved one's home or business.

— **THOMAS P. HUGHES**, *Networks of Power*, 227

> ### *Electricity in New York City, 1882*
>
> The whole lamp looks so much like a gas burner surmounted by a shade that nine people out of ten would not have known the rooms were lighted by electricity, except that the light was more brilliant than gas and a hundred times steadier. To turn on the light nothing is required but to turn the thumbscrew; no matches are needed, no patent appliances. As soon as it is dark enough to need artificial light, you turn the thumbscrew and the light is there, with no nauseous smell, no flicker and no glare. . . .
>
> The light was soft, mellow, and grateful to the eye, and it seemed almost like writing by daylight to have a light without a particle of flicker and with scarcely any heat to make the head ache. The electric lamps in THE TIMES Building were as thoroughly tested in a single evening, and tested by men who have battered their eyes sufficiently by years of night work to know the good and bad points of a lamp, and the decision was unanimously in favor of the Edison electric lamp as against gas.
>
> – *New York Times*, 5 September 1882

of electric storage batteries in the early 1800s. Inquisitive researchers continued to build more complex equipment and to imagine other possible uses for electricity. By the 1830s, they had developed rudimentary electrical motors capable of real work.[2]

The nineteenth-century public was fascinated by electrical motors but few people imagined that such devices would ever seriously compete with the steam engine. The ultimate symbol of the industrial revolution, the steam engine was a versatile and reliable source of power. In 1869, for example, locomotives powered by steam began crossing North America, proof of the utility of the steam engine. By contrast, the crude experiments with electricity seemed little more than a novelty.

Advances in lighting during the 1870s and 1880s helped to make electricity a common feature of North American life. Thomas Edison's well-publicized incandescent light in 1879, in particular, and his role in building a central power station in New York City over the next three years, bolstered electricity's public image considerably.[3] Edison himself was an extraordinary individual, possessing almost as much entrepreneurial talent as he did inventive genius. Shortly after construction had begun on his power station, for example, he invited all of New York's aldermen out to his New Jersey research laboratory. They made the trip on a dark December evening in 1880, so he thoughtfully provided electric lights to light their path as well as a sumptuous feast catered by Delmonico's. He needed city approval to dig up the streets around the power station (situated in the heart of New York); the aldermen rose after dinner to give him their hearty approval. Edison was much in the public eye in the late nineteenth century, forever showing off gadgets and promising new wonders.

Canadian cities were quick to adopt electricity. Although one journalist concluded in 1884 that there was "a decided impression in the popular mind of Toronto that electricity as a street illuminator is a failure," five years later the city had some five miles of underground electric cable. By the end of the 1880s, most Canadian cities of any size had electric street lighting and large American electrical manufacturers had begun to set up shop in Quebec and Ontario.[4]

Ottawa, 1893

Since 1885 the introduction of electricity has been general, and to such an extent that almost every shop in town, and hundreds of private dwellings, now enjoy the luxury of its soft white light.

— **FRANK MCDOUGAL**, "Electricity in Ottawa," *The Owl*, February 1893

The practical application of electricity grew by leaps and bounds during the period from 1885 to 1895. Entrepreneurs and inventors worked feverishly, promoting their own patented products. Frequently these innovations were mutually incompatible. The most famous example of this was the bitter rivalry between the advocates of direct current and alternating current (commonly known as "the battle of the currents"). Edison, although he remained electricity's foremost advocate, clung to his initial faith in direct current, while another canny inventor and business person, George Westinghouse, marketed a competing alternating current system. Edison failed to recognise — or chose to ignore — the growing advantages of alternating current. AC power was not only better suited for transmission than direct current but also had a decided cost advantage. This did not stop Edison from opposing the use of alternating current, an opposition that would take bizarre twists.

One of the advantages of direct current, Edison and his colleagues claimed, was that it was far less dangerous than alternating current. They threw their considerable weight behind a public campaign aimed at curtailing AC power by limiting electrical transmissions to a maximum of 300 volts. One gruesome aspect of such public posturing was the subtle and not so subtle promotion of the electric chair as a means of public execution.[5]

One business person with links to the Edison group claimed darkly that "strenuous attempts have been made to befog the public mind in order to prevent the use of alternating current for the death-penalty, lest the public should learn its deadly nature and demand that the Legislature banish it from streets and buildings, thus ending the terrible, needless slaughter of unoffending men."[6] As this passage hints, New York state had adopted the electric chair in 1888 as a means of execution, opting to use alternating current to produce the lethal shock.

Death by any other name

In the courtyard of [Thomas Edison's] West Orange Laboratory, media-event public executions of animals by alternating current were conducted [in the late 1880s]. . . . Edison's attorneys Eaton and Lewis suggested in all seriousness that recommended terms for this method (which included "ampermort," "dynamort," and "electromort") were inaccurate in their derivation: "Electricide" seemed more fitting, expressing direct analogies in the English language with "homicide" and "suicide."

But perhaps the best choice of all was to invent a new verb, "to *westinghouse*. . . . As Westinghouse's dynamo is going to be used for the purpose of executing criminals," Edison's cunning men reasoned, "why not give him the benefit of this fact in the minds of the public, and speak hereafter of a criminal being 'westinghoused,'" or (to use it as a noun) "as having been *condemned to the westinghouse*" . . . ?

— **NEIL BALDWIN**, *Edison: Inventing the Century*, 202

Edison ultimately had to admit defeat. Alternating current soon became the most common form of electricity produced by public utilities, largely because AC power could be transmitted efficiently over relatively large distances. By contrast, direct current could be transmitted no more than a mile or so. However, this was by no means the only factor in AC's growing domination of the electrical field. The price of copper had climbed in the late 1880s, a factor which increased the cost-effectiveness of AC, since its transmission lines could be kept much thinner than those which carried DC power. In addition, a former Edison employee developed an efficient AC motor in 1888. This was a real breakthrough for AC power since up to this point the lack of such a motor had held back its growth. Alternating current also received a good deal of favourable publicity in the early 1890s. The organisers of the 1893 Chicago world's fair — known as the Columbian Exposition — opted to use AC power, as did the company that was building an elaborate and much-discussed hydro-electric project at Niagara Falls.[7]

The Chicago Columbian Exposition of 1893 had many purposes. Business people in the city were anxious to shed Chicago's image as a place torn by bitter strife between working people and the authorities, a reputation it had earned with the Haymarket Riots of 1886. More generally, the fair commemorated the four hundredth anniversary of

"The Edison Electric Tower" at Chicago's Columbian Exposition, 1893. Visitors were impressed, as the following description suggests: "When, at the silent touch of an unseen hand, the tower from base to apex is arrayed in robes of scintillating and many coloured lights, we have here the very incarnation of electrical science." — Hubert Howe Bancroft, *The Book of the Fair: An Historical and Descriptive Presentation of the World's Science, Art, and Industry, as viewed through the Columbian Exposition at Chicago in 1893*. Chicago: Bancroft, 1893, 406.

The Electricity Hall at the Columbian Exposition. The fair's lights at night "brought visitors to the grounds by scores of thousands. . . . Under the glow of myriads of lamps the architectural symmetry of the design is displayed to excellent advantage, the uniform cornice level of the buildings, sixty feet above the ground, standing clearly forth in tracery of incandescent lights, while encircling the basin near its water level an unbroken circle of vari-colored jets, each reflected from its glittering surface, gives to the scene a brilliance almost too dazzling for human eyes to rest upon." —*Ibid.*, 402.

Columbus's landfall in North America. But the star of the fair—quite literally, at night—was electricity.[8] Well over twenty million visitors came to the fair and none could avoid the many displays of electricity. At night electric lights illuminated the buildings and shot rays up into the dark sky, while during the day electricity transported visitors along a moving sidewalk, on elevators within many of the fair buildings, on a railway that ran along the fairgrounds, and in any one of the nearly sixty battery-powered boats on the water.[9] The Electricity Building itself covered more than five acres, its spacious interior dominated by the huge "Edison Electric Tower." But by this time Edison had moved on to other ventures. The year before the Chicago fair, his company, Edison General Electric, merged with a rival to form General Electric. The dropping of Edison's name was more than a symbolic gesture and he himself was little interested in the emerging corporate giant.[10] General Electric also acknowledged the advantages of alternating current and began producing its own line of AC generators.

Canadians, then as now, followed events in the United States closely. Electricity — especially hydro-electric developments — seemed well-suited to a country with such a range of river systems, with such long dark winter nights.[11] Coal, the common source of energy and heat in nineteenth century Canada, was not without disadvantages. It was relatively expensive and supplies could be disrupted by bitter labour disputes. By contrast, electricity appeared to promise a stable and inexpensive alternative. Little wonder that it was embraced with enthusiasm by business people and consumers alike. As in the United States, enthusiastic supporters of electricity staged rather unusual displays to impress the sceptical. One of the more unique demonstrations occurred at a banquet in 1882. Bowls with various hues of coloured water in which fish swam decorated the tables, illuminated by electric lights.

A Very Beautiful Effect

At the junior conservative ball held in St. Lawrence Hall in 1882, at which Lady Macdonald was Lady Patroness, we had placed upon the banquet tables a number of fish bowls, some with different coloured waters in them, and others with clear water, and gold and silver fish therein. We had incandescent lamps placed in them, and when the guests sat down to the tables, we switched in the current, and I can assure you it had a very beautiful effect. We had to keep ice constantly in the water to keep it from overheating. . . .

— **JOHN SMILLIE,**
Canadian Electrical News and Steam Engineering Journal, January 1894

Electricity came to British Columbia in the early 1880s. People in Victoria had grown dissatisfied with the street lighting provided by the local gas company and so were prepared to experiment with electricity. Although its success was debatable — soon there were as many complaints about the electric lights as the earlier gas lamps — it was an important beginning.[12] As with street lighting plants for other Canadian cities, Victoria received its electrical equipment from south of the border, from the Brush Company in San Francisco. The great moment came just before Christmas in 1883 when bright arc lights atop three towers in Victoria's downtown district lit up the night for the first time.

Electricity also powered Victoria's tramway system, which began its service in early 1890. Both the city's street lighting and its trams had their share of financial problems, compounded by plain bad luck.

> **Southern Exposure**
>
> The telephone, electric light, and trolley took different routes into Canada, but they all came from the same place, entering the main Canadian cities in the first wave of diffusion from centers of innovation in the northeastern United States. Canada became tied into a continental technological network despite its strong political and economic connections with Great Britain; it was US rather than European styles, standards, and systems that were transferred to Canada.
>
> — **CHRISTOPHER ARMSTRONG** and **H. V. NELLES**, *Monopoly's Moment*, 91

In 1885, two years after beginning operations, the private company that supplied Victoria with street lights was bought out by the city of Victoria. The tramway company also tried—unsuccessfully—to interest the city in taking over its assets. After a series of calamities, the company went into receivership in 1895. On the mainland, the pioneer tramway companies of both Vancouver and New Westminster were just as unsuccessful. But these business failures led to the creation of one of the province's most successful companies, the London-based British Columbia Electric Railway Company Limited. This company was formed in the spring of 1897 and almost immediately acquired the assets of the defunct tramways of Victoria, Vancouver and New Westminster.[13] More importantly, the new company also became the lower mainland's major electrical utility. Its dominance lasted until 1961, when the province took it over and created BC Hydro.

The early experiments with electricity in the province were not all that successful. Their significance lies in the fact that they set the stage for the remarkable growth that occurred in the early twentieth century. The completion of Canada's first transcontinental railway in 1885 was a much more profound event, for it altered the province's economic development in important ways. Victoria's merchant elite, the dominant group in both politics and business, faced new challenges from central Canada. The new city of Vancouver, in almost every way an offshoot of the railway, gradually eclipsed the older centres of Victoria and New Westminster. The grandiose new parliament buildings, begun in 1893 and completed five years later, stand as the last tangible symbol of Victoria's prestige and influence. The new century was to belong to Vancouver.

As trains began to cross British Columbia, the mountains separating the province from the rest of Canada lost much of their significance. Geography became less of a factor when travelling: people on the coast

BCARS B-05160

A CPR train pauses above West Kootenay Power's plant on the Kootenay River, some time in the spring of 1898. These two icons of late nineteenth century technology — the steam locomotive and hydro-electricity — came in the wake of the first mineral discoveries in the region.

could catch a train to Kamloops or Revelstoke and then take a steamer along the rivers and lakes of the southern Interior. And in the 1890s the region attracted many people. A mining boom was underway in the Kootenays, creating cities in an area that had once seemed isolated and remote.

Like the railway to the Pacific coast, the mining industry introduced the latest features of North American technology to the Kootenays. The increasing pace of industrialization acted as a spur to mining: the growing demand for metals such as copper, lead, and zinc reflected the rapid advances in technology typical of the late nineteenth century. The work of mining and the various methods of ore treatment could only be carried out if an adequate supply of power was available. Engines driven by steam or electricity were an essential part of the industrial complexes that emerged in the Kootenays by the late 1890s. West Kootenay Power was a product not only of the advances in applied technology pioneered by people such as Edison, Westinghouse, and others in the 1870 and 1880s but also of the specific energy requirements of the region in which it was formed.

The Spirit of the Age

The railroad as a system incorporated most of the essential features of the emerging industrial order: the substitution of metal for wood construction; mechanized motive power; vastly enlarged geographical scale; speed, rationality, impersonality, and a spirit of efficiency that included an unprecedented emphasis on precise timing.

— LEO MARX, *New York Review of Books*, 15 March 1984, 28

Business people involved in Rossland's booming mining industry formed the West Kootenay Power & Light Company Limited in 1897. This was the town's heyday the mines of Rossland were famous. Newspapers in Toronto and Montreal carried almost daily reports on the area in 1896 and 1897 (two of them produced special issues on BC's mining industry[14]). Door-to-door sales people in Ontario sold gold mining stocks along with their other wares while clergymen and professors publicly condemned the folly of the gold craze.[15] In Rossland itself, managers rushed development, anxious to ship ore as soon as possible. West Kootenay Power & Light came out of this feverish atmosphere, created to supply power to one of Rossland's leading mines, the Centre Star.[16]

The company was not the first to provide electricity for the community of Rossland. In 1895, two years prior to West Kootenay's formation, several American business people organized the Rossland Water and Light Company. As the name suggests, this utility was to supply the town with both electric light and water. Pure water was particularly important. Like other instant towns, Rossland at first found it difficult to supply its mushrooming population with such necessities. The consequences were predictable: the secretary of the Provincial Board of Health blamed "the unsanitary condition of Rossland in 1895 and 1896" for the three hundred cases of typhoid fever reported during those two years.[17] Everyone welcomed the company's plans to pipe water from Stoney Creek into town. Once this was done, people would no longer have to depend solely on the limited and often contaminated water supply available within Rossland itself. During the winter of 1895–96, crews built a twelve-inch-square wooden flume from the creek into a large storage tank above the centre of Rossland.

Work on the electrical plant went ahead hand in hand with the water works. The company set up a steam engine to drive a generator. The local newspaper reported that on the evening of 7 January 1896 the first electric lights shone in Rossland. The town's growth meant that the pioneer light and water company was soon struggling to keep up with the demand for both the commodities it sold.

The Lights of Rossland

The few buildings which had been equipped with incandescent lamps looked very brilliant in contrast with their neighbours. Mrs. Allen's hotel was a blaze of light from basement to attic and Stack & McDonald's saloon looked fairly palatial. Almost every house on Columbia avenue, Spokane street and "Sour Dough Alley" has been wired for lights, but the supply of lamps on hand was altogether inadequate and another consignment will have to be received before the public demand is satisfied.

– *Rossland Miner*, 11 January 1896

People welcomed the appearance of the Rossland Water and Light Company and the vital public services that it undertook to provide. Unfortunately for Rossland, the company's owners were less interested in the community's long-term growth than they were in profiting from some well-timed investments. Their most valuable asset was not the Rossland Water and Light Company but the War Eagle mine, one of Rossland's leading properties. The mid-1890s mining boom had driven up the value of the War Eagle. In December 1896 its owners seized the opportunity to sell the mine to a Toronto syndicate for $825,000. Pocketing their windfall profits, they returned to Spokane.[18] Rossland's need for energy remained, however. A reliable and abundant supply was essential for profitable mining operations, the economic activity upon which the community was based.

Hydro-electric projects were a novelty in the 1890s but a growing number of engineers were interested in exploring their potential use in industry. Engineers were an occupational group well-represented in Rossland since most of the town's many mining companies employed at least one. Some — if not all — of these men would have been familiar with the successful experiments in generating hydro-electricity.[19] The nearby Columbia and Kootenay rivers suggested that a cheap and effective solution to the mines' constant power demands was at hand.

NOTES

1 These statistics are provided in Thomas P. Hughes, *Networks of Power: Electrification in Western Society, 1880–1930* (Baltimore: Johns Hopkins University Press, 1983), 227.

2 W. James King, *The Development of Electrical Technology in the 19th Century* (Washington, DC: US Government Printing Office, 1961, Bulletin 228, Paper 28), 260-69.

3 The best book on Edison is Neil Baldwin's *Edison: Inventing the Century* (New York: Hyperion, 1995), although the brief pamphlet by Hughes puts much of the important detail onto fewer pages: Thomas P. Hughes, *Thomas Edison: Professional Inventor* (London: HMSO, 1976). Other useful works include Robert D. Friedel and Paul Israel, with Bernard S. Finn, *Edison's Electric Light: Biography of an Invention* (New Brunswick, NJ: Rutgers University Press, 1986), and William S. Pretzer, ed., *Working at Inventing: Thomas A. Edison and the Menlo Park Experience* (Dearborn, Michigan: Henry Ford Museum & Greenfield Village, 1989).

4 The quotation in the text is from "Here and There," *The Week*, Vol. 1, No. 38 (21 August 1884): 596. For accounts of American plants being established in Canada, see "The Edison General Electric Company, Sherbrooke," *The Dominion Illustrated*, Vol. 5, No. 113 (30 August 1890): 155, and the description of the Edison works in Peterborough, Ontario, in the *Canadian Electrical News and Steam Engineering Journal*, Vol. 1, No. 10 (October 1891): 135. The Sherbrooke plant was begun in 1889, the Peterborough plant in 1888.

5 See especially the account by Thomas Hughes, "Harold P. Brown and the Executioner's Current: an Incident in the AC-DC Controversy," *Business History Review*, Vol. 32, No. 2 (1958): 143-65; also, Terry S. Reynolds and Theodore Bernstein, "The Damnable Alternating Current," *Proceedings of the IEEE*, Vol. 64, No. 9 (September 1976): 1339-43 and Edison's own intervention, Thomas A. Edison, "The Dangers of Electric Lighting," *North American Review*, Vol. 149, No. 396 (November 1889): 625-34.

6 Harold P. Brown, "The New Instrument of Execution," *North American Review*, Vol. 149, No. 396 (November 1889): 593.

7 See Robert Belfield, "The Niagara System: The Evolution of an Electric Power Complex at Niagara Falls, 1883–1896," *Proceedings of the IEEE*, Vol. 64, No. 9 (September 1976): 1344-50, esp. 1348.

8 See the discussions of the Chicago Fair in Baldwin, *Edison: Inventing the Century*, 233-38; Carolyn Marvin, *When Old Technologies Were New: Thinking about Electric Communication in the late Nineteenth Century* (Oxford: Oxford University Press, 1988), 171-73; David E. Nye, *Electrifying America: Social Meanings of a New Technology, 1880–1940* (Cambridge, MA: MIT Press, 1990), 37-41; James Gilbert, *Perfect Cities: Chicago's Utopias of 1893* (Chicago: University of Chicago Press, 1991), 75-130; and Robert W. Rydell, *All the World's a Fair: Visions of Empire at American International Expositions, 1876–1916* (Chicago: University of Chicago Press, 1984), 38-71. For some insightful

comments on the broader role and meaning of such fairs, see Paul Greenhalgh, *Ephemeral vistas: The Expositions Universelles, Great Exhibitions and World's Fairs, 1851–1939* (Manchester: Manchester University Press, 1988), *passim*.

9 David E. Nye, *Electrifying America*, 38.

10 See the account in Baldwin, *Edison: Inventing the Century*, 227-29; Alfred D. Chandler, Jr., *The Visible Hand: The Managerial Revolution in American Business* (Cambridge, MA: Belknap Press, 1977), 426-33; and Thomas P. Hughes, *Networks of Power*, 106-39.

11 See especially the comments of A. J. Lawson, "Generation, Distribution and Measurement of Electricity for Light and Power; Appliances Therefor, and Particulars of Canadian Installations," *Transactions of the Canadian Society of Civil Engineers*, Vol. 4 (1890): 179-240, and T. C. Keefer, "Presidential Address," Royal Society of Canada, *Proceedings and Transactions*, Vol. V (1899): 3-40.

12 See Patricia Roy, "The Illumination of Victoria: Late Nineteenth-Century Technology and Municipal Enterprise," *BC Studies*, #32 (Winter 1976–77): 79-92; G. W. S. Brooks, "Edgar Crow Baker: An Entrepreneur in Early British Columbia," *BC Studies*, #31 (Autumn 1976): 23-43; and Douglas V. Parker, *No Horsecars In Paradise: A History of the Street Railways and Public Utilities in Victoria, British Columbia before 1897* (Toronto: Railfare Enterprises, 1981), 10-19.

13 See Parker, *No Horsecars In Paradise*, 100-24, and Patricia E. Roy, "The British Columbia Electric Railway Company, 1897-1928: A British Company in British Columbia," PhD thesis, University of British Columbia, 1970, 1-43.

14 *The Evening Star* (Montreal), 19 January 1897; *The Saturday Globe* (Toronto), 6 February 1897.

15 See "Notes on Canada's Mines," *The Globe*, 24 December 1896, 9; "The Mines of Canada," *The Globe*, 9 February 1897, 5; "Mining Development," *The Globe*, 15 February 1897, 12. One journal despaired of "the wave of partial insanity, which has broken loose in Toronto and threatens to attack Montreal" ("Truth vs. Exaggeration," *Canadian Mining Review*, Vol. XV, No. 12 (December 1896): 248).

16 "The primary purpose of the undertaking is to provide power for the working of the Centre Star mine" ("A Big Power Plant," *Rossland Miner*, 6 July 1897, 3); see also "From Western Contemporaries," *The British Columbia Review*, 4 September 1897, 200.

17 "Annual Report of the Provincial Board of Health," British Columbia, *Sessional Papers*, 1898, 1165.

18 "The Sale Assured," *Rossland Miner*, 7 January 1897, 3.

19 The Niagara Falls hydro-electrical development attracted much publicity; more technical work appeared in various engineering journals. See for example Charles F. Scott, "Long Distance Transmission for Lighting and Power," *Transactions of the American Institute of Electrical Engineers*, Vol. 9 (7 June 1892): 425-44, although even popular magazines carried news of these developments (for example, John McGhie, "Long-Distance Transmission of Power by Electricity in the United States," *Cassier's Magazine*, Vol. 9 [February 1896]: 359-74). Mining engineers would also have known of the early success with hydro-electricity at Telluride Colorado, which began producing power in late 1890. According to P. N. Nunn, this was "the first commercial, high-pressure, alternating-current power transmission of the world" ("Pioneer Work in High-Tension Electric Power Transmission: The Operations of the Telluride Power Company," *Cassier's Magazine*, Vol. 27, No. 3 [January 1905]: 171-200).

CHAPTER TWO

Starting Out: *The First Years of West Kootenay Power & Light*

In the spring of 1897 three business people petitioned the BC legislature, requesting permission to establish a hydro-electric power plant and form a business to be called the West Kootenay Power & Light Company, Limited. It's worth noting that the three — Oliver Durant and Patrick Largey from south of the border and C. R. Hosmer from Montreal — were all self-made men, entrepreneurs rather than technical specialists with a particular interest in electrical generation. They launched West Kootenay Power because it seemed a promising business venture.

The proposed West Kootenay plant was an idea whose time had come. That year ten other companies also sought the provincial government's approval to supply power to various communities and regions in the interior of the province.[1] Most were speculative ventures and lacked the solid financial support that lay behind West Kootenay Power. When the legislature gave final approval to the West Kootenay bill on 8 May 1897, the company's promoters immediately began the necessary work involved in the construction of a large and complex enterprise. Some fourteen months later, on 15 July 1898, electrical current passed through the transmission lines from the company's generators below the dam on the Kootenay River to the community of Rossland, some thirty miles away.

With hindsight, it seems obvious: the falls on the Kootenay River were ideally suited for power generation and the communities of the region — especially Rossland with its booming mines — needed power. The sponsors of West Kootenay Power did little more than seize a favourable business opportunity. But success demanded more than calculating the simple arithmetic of investment and likely profits. The number of applications to incorporate power companies in 1897 indicates that others had also worked out this attractive equation and reckoned on the dividends awaiting those who could deliver electricity to the Rossland mines. Few applicants, however, possessed the necessary business acumen, access to credit, and engineering skill to carry a large project through to the end. These were the essential ingredients that ensured West Kootenay Power's success. They were human attributes and, naturally enough, belonged to people rather than a newly-incorporated business.

The biographies of the three individuals who created West Kootenay Power tell us something of the nature of business in the Kootenays during that era. The two Americans, Patrick Largey and Oliver Durant, were representative of the entrepreneurs chiefly responsible for the early development of Rossland's mines. This was a relatively homogenous group. Most had moved to the Rocky Mountain region as adults, and a number of them had participated in the Idaho mining boom of the 1880s before turning their sights north to the mines of British Columbia. Their interest in Rossland was to prove fleeting; few of them lived there for any length of time and all took advantage of the opportunity to sell their properties once the price was high enough.[2]

Patrick Largey, President of the Montana-based Centre Star Mining & Smelting Company, was the nominal head of the drive that spring in 1897 to incorporate West Kootenay Power & Light. Born in Ohio, Largey had come west as a young man in 1863. By the 1890s he was a prosperous business person in Butte Montana, the president of one of the city's banks as well as a large hardware store. A disastrous event occurred at this store in January 1895: a fire detonated explosives held in the store, causing a blast which killed fifty-six people. Largey soon got over this setback and subsequently invested heavily in Rossland. He was best-known in the Kootenays for his involvement with the Centre Star mine, of which he was president; but in Butte, his past was not forgotten. Thomas Riley had lost family as well as a leg in the hardware blast, and the memory of that terrible event still burned in him. In January 1898, as West Kootenay's construction crews worked as hard as weather conditions would permit, Riley hobbled into Largey's bank in Butte and pulled out a revolver. His second shot killed Largey.[3]

Oliver Durant counted Largey as one of his dearest friends. They were close in age — near sixty when the company was formed in 1897 — and their lives had followed a similar pattern. Like Largey, Durant moved west from his native New York at a young age and by the 1890s was a successful business man.[4] He profited from the Idaho mining boom in the 1880s as well as the spectacular rise of Rossland the following decade. When West Kootenay Power & Light was formed, Durant was the Centre Star's general manager. For two years — from 1897 to 1899 — he served as West Kootenay Power's first president.

C. R. Hosmer was the third member of the group and like Largey and Durant was very much a self-made man. He had risen to prominence as manager of the Canadian Pacific Railway's telegraph system. After ten years in this job, however, Hosmer became restless. In his mid-forties he was drawn, like so many others, to Rossland. He invested in real estate and mines as well as the power company, and in 1899 left the CPR to give his undivided attention to these new interests. Later, in 1913, he was to become president of West Kootenay Power & Light. By this time he was a leading member of the Canadian business elite: one publication identified him as one of the twenty-three men "at the basis of Canadian finance."[5] As well as being president of West Kootenay Power, he was a director of a handful of Canada's most important companies, including the Bank of Montreal, Dominion Textile, Laurentide Paper, Consolidated Mining and Smelting, London and Lancashire Life, Sun Life Assurance, and the Royal Trust Company. Hosmer's contacts with other members of the central Canadian establishment helped to determine West Kootenay Power's future.

The relationships between Largey, Durant, and Hosmer were no doubt complex and really we can only speculate about the content of their conversations, their individual motives, and how they arrived at the decision to form the company. Enough evidence survives, however, to sketch at least the broad details of that last event. In fact, West Kootenay Power & Light began as not one but two companies. Durant, Largey, and Hosmer signed the original petition for legislation to form the West Kootenay Power & Light Company, seeking permission "to appropriate and use waters from Sheep Creek." But others submitted a second petition with the same objective, to provide power for the region's mining industry, using water "from Kootenay River and its tributaries, and the Columbia River and its tributaries, as the Company may see fit."[6]

This second company included two men destined to play a key role in the future of West Kootenay Power. One was an influential member of the Montreal business community. William Doull was close to fifty when West Kootenay Power was launched and came from a prominent Halifax family (his father "one of the great merchant princes of the Nova Scotia capital"). Doull however had opted to leave Halifax for Montreal. At first he ran a branch of the family's clothing firm in his adopted home but soon began to diversify, investing in a number of

ROSSLAND HISTORICAL MUSEUM

By 1897 the possibility of generating power from the Kootenays' rivers was attracting much attention. This contemporary photograph shows the Sheep Creek Falls, originally intended as the power site for West Kootenay Power & Light Company.

power companies. The West Kootenay Power & Light was closest to his heart but he was also interested in similar companies in Mexico and Cuba.[7] Although he did not become president of West Kootenay until 1903, the company's early files make it clear that from the first Doull figured prominently in its operations. He was still president in 1913 when he passed away.

Sir Charles Ross was the other person associated with the second proposal to develop the hydro-electric potential of the Kootenay River. He was, to put it mildly, an eccentric individual — although to give him his due, he was widely acknowledged as the person most responsible for pushing the construction of the West Kootenay project. A British aristocrat, Ross arrived in Canada in 1896, just twenty-four years old and not long out of Cambridge. Reputedly the largest land-owner in Scotland, he was then in the middle of an acrimonious divorce from his wife of three years, an event which likely accounts for his trip across the

Atlantic. In Montreal, Ross made Hosmer's acquaintance, and Hosmer took the trouble to get his new friend an engine pass as he set out across the country for the Kootenays.[8] On the lookout for business opportunities, Ross bought into Largey and Durant's Centre Star mine and was subsequently made a director of the company.[9] Ross succeeded Durant as president of West Kootenay Power, a position he held for four years, from 1899 to 1903.

Sir Charles Ross and Oliver Durant managed to bring together these two groups: Largey, Durant, and Hosmer, the promoters of West Kootenay Power & Light, and Doull, Ross, and several others, who were behind the similarly-named Kootenay Power & Light. The result of this union was the final version of West Kootenay Power & Light. A newspaper story published a dozen years later suggested that an evening of drinking led to the union. Durant was serving up Scotch to three friends—including Ross—at the office of the Centre Star mine

WEST KOOTENAY POWER – TRAIL OFFICE

Men work to raise a double pole for the power line across the Columbia River, using rope, blocks, and a lever. The photograph shows the poles' cedar caps, intended to keep the snow off the insulators. It soon turned out that the caps were unnecessary, but they proved a favourite haunt for birds. "A nightmare for 20 years," grumbled one man, who routinely had to remove electrocuted birds from the lines.

when the conversation turned to water power. At the time, Durant was involved in a dispute over water rights on Sheep Creek, six miles to the west of Rossland, likely because of his petition then before the legislature. One of Durant's guests suggested that Sheep Creek was a poor choice if Durant was seriously interested in power generation. The falls on the Kootenay River west of the city of Nelson, the man pointed out, would be a much better choice for a hydro-electric plant. Durant and Ross decided to investigate and arranged to pay a visit to the area.[10] Ross named the Kootenay's cascading waters the Bonnington falls, commemorating his Scottish homeland. At this point, apparently, the two men decided to work together to forge the company that would harness the abundant water power of the falls.

After the two groups that had petitioned the legislature joined forces, Ross assumed the task of supervising construction. In early May 1897, a couple of days before the Legislature formally assented to the West Kootenay bill, Ross entered into a formal agreement with his colleagues. He would build the Bonnington plant, erect the transmission lines, and so on, in exchange for $125,000. Once this task was completed, Ross would turn everything over to the company and in exchange he would receive a substantial amount of company stock as well as bonds. Contemporary newspaper accounts recognized Ross's central role in the project. When the *Montreal Gazette* interviewed Ross in late 1897, for example, it noted that he was "well known in connection with the development of Rossland . . . , besides being the moving spirit in the West Kootenay Power & Light Company."[11]

Ross had studied mechanical engineering at Trinity College, Cambridge, and as a result was well-prepared to assume the role of general contractor for the West Kootenay project. Surviving letters from the period make it clear that he took the job of supervision seriously. He kept in frequent touch with the Toronto-based Canadian General Electric, the company which supplied virtually all of the electrical equipment. These letters included sketches, questions, and suggestions for improvements and modifications to the equipment.[12] His contact at Canadian General Electric was that company's chief engineer, Lorne Campbell. Ross soon came to trust Campbell and in January 1898 announced that he was putting Campbell in charge of the project. Other projects demanded Ross's attention.

Ross was an unlikely business executive. Although capable of flashes of inventive genius, he had little time for the necessary and routine tasks that provide stability and direction to a successful company. Oliver Durant, his predecessor as president of West Kootenay,

ROSSLAND HISTORICAL MUSEUM

West Kootenay Power & Light advertises its product: blazing lights illuminate the building housing the company's office (note the sign in the bottom left window).

confided some years later to a newspaper reporter that Ross "was a source of considerable grief and anxiety in many directions."[13] During the four years that Ross was president of the company, he spent time just about everywhere in the world except British Columbia. In 1898 and 1899 he was negotiating with the Russian and Chinese governments, hoping to build a railway in Manchuria. In 1900 he was off to South Africa, where he spent six months fighting in the Anglo-Boer War. The following year he married for the second time, to a young American woman that he had met in London. Increasingly, however, rifle-manufacturing took up his time and attention. He had set up a small factory in Connecticut to produce rifles in 1896. After negotiating an agreement with the Canadian government in 1901–1902—undertaking to supply 10,000 rifles annually for twenty-five dollars each—Ross moved his factory from New England to Quebec City. (Subsequently the standard issue for Canadian soldiers, the Ross rifle was embroiled in controversy during the First World War.[14]) An avid sportsman, he also returned regularly to his Scottish estate, which boasted 60,000 acres of grouse moors, 100,000 acres of deer forests, as well as excellent salmon and trout fishing. From 1899 to 1903, the years he was president of West Kootenay Power & Light, Ross could not have spent much time attending to the company's affairs.[15]

Fortunately for West Kootenay Power & Light, Ross made an excellent choice when he picked Lorne Campbell to manage the company's affairs. At first Campbell simply took over the task of supervision but in April 1898 he left his job with Canadian General Electric and became general manager of West Kootenay. Right up until his death nearly fifty years later, Campbell was intimately associated with the company's affairs.

Raised and educated in Ontario, Campbell joined the engineering staff of the Edison General Electric Company in Toronto in 1889 when he was still in his teens. Two years later he became chief engineer of its successor company, Canadian General Electric. Ambitious and hard-working, Campbell was twenty-seven when he left his job with Canadian General Electric to become general manager of West Kootenay Power. His first task with the new company was to see the construction phase through to a successful conclusion. On the first of June 1898

Producing Power

The plant at Bonnington is now in operation running on a water load for the past two weeks.

For the past week we have been transmitting 1,000 HP to Rossland; the last three days of the week we were transmitting this under a pressure of 20,000 volts. Last night we operated for 8 hours through a perfect downpour of rain accompanied by sharp lightning, and experienced no trouble. . . .

So far the electric plant and water wheels are giving entire satisfaction, and we are tonight running 500 lights for the Rossland Water and Light Company. Tomorrow night we expect to have the greater part of the commercial lighting cut in on our circuits and from this date forward, the plant will be in continuous operation. I might mention that it is exactly one year and one day since the installation of this plant was commenced.

— L. A. CAMPBELL to C. R. HOSMER, 15 July 1898

Campbell wrote to William Doull in Montreal, assuring him that "no time will be lost in putting the plant in operation and am pushing things to completion with the utmost possible speed."[16] Six weeks later he proudly reported that electricity was being transmitted to Rossland without a hitch.

Just as the first Bonnington plant began to produce power, Oliver Durant and others involved with the Centre Star mine decided that the time was ripe to sell. An influential Toronto family group—the Gooderham liquor interests—purchased the property in September 1898 for two million dollars.[17] The sale of the mine combined with ill health encouraged Durant to consider retirement. Like other successful American investors in Rossland, he returned south to a handsome mansion in Spokane. By 1900 the company's first president no longer owned shares in West Kootenay.[18]

The year 1898 marked a watershed of sorts for the Kootenays. American ownership and influence declined, and more and more Canadians became interested in the region. Two famous Rossland mines—the Le

BCARS B-05157

Two fishermen try their luck in the tail race of the recently-completed power house at Lower Bonnington.

Roi as well as the Centre Star — passed out of American hands during the year. In another important transaction — one which was to have a profound effect on the future of West Kootenay Power & Light — Canada's most powerful business group, the Canadian Pacific Railway, acquired the Trail smelter and constructed a branch line through the Crowsnest Pass into the region.

And West Kootenay Power & Light was beginning to attract attention. These were, after all, the pioneer days of hydro projects. BC Electric's Goldstream plant on Vancouver Island — which began generating power in September 1898 — was the first such project out on the coast and was a much less ambitious facility than West Kootenay Power's Bonnington plant. Electricity was rapidly emerging as the technology of the future, much discussed and associated with optimistic expectations of a brave new world. Up-state New York, for example, enjoyed a dynamic period of growth after the first hydro-electric plant was installed on the American side of Niagara Falls and was frequently mentioned as an example of the benefits of hydro-electricity.[19]

A Big Power Plant

The primary purpose of the undertaking is to provide power for the working of the Centre Star mine, in which Sir Charles Ross holds a considerable interest, but the scope has been much broadened, and it is now the purpose of the company to establish a plant which will generate electric power to be used all over West Kootenay. . . .

The change will work as great a revolution in this camp as the introduction of Niagara electric power is working in the city of Buffalo, NY. . . .

The putting in of [West Kootenay Power's] great electric plant will not at first be understood. It means a revolution in the expense of mining in Rossland. . . . The plant will be of the most improved construction and will be one of the finest of the kind in the world. . . .

The commencement of work on this enterprise marks an epoch in the history of the Rossland camp, and of West Kootenay.

– *Rossland Miner*, 6 July 1897

The success of the West Kootenay plant meant that it too became an example of the way of the future. Newspapers and magazines described West Kootenay's operations in glowing terms, coverage that extended well beyond the local press. Canada's leading electrical magazine, the *Canadian Electrical News and Steam Engineering Journal*, carried

Number One plant at Lower Bonnington. Above, the simplicity and perspective of this carefully-planned photograph suggest the interior of a church as much as that of an industrial plant. Right, operators pose outside the building. Ties, hats, and moustaches were all evidently standard items.

numerous articles on West Kootenay Power's progress throughout 1897 and 1898, for example, and in mid 1899 the San Francisco-based *Journal of Electricity* (sub-titled "An Illustrated Review of the Industrial Applications of Electricity, Gas and Power") carried a two-part feature article on the company. The magazine heaped praise on its plant, judging it "one of the most perfect mining transmissions to be found on the Pacific Coast."[20] The article described the power house, the dam, transmission lines, and the Rossland sub-station in glowing terms but acknowledged that "the principal interest in the Kootenay-Rossland transmission centers in its application of electric power for mining and milling purposes."[21] The article then offered a lengthy description of the electric hoisting plant of the War Eagle mine, driven by a 300 horse power induction motor. As well as text, the piece included a number of line drawings and photographs of the hoist's equipment. The author could not hide his fascination with the equipment, confessing to his readers that "one who is interested could spend hours in watching the operation of this hoist."[22] Nor was he alone. That same year, Victoria's *Mining Record* carried an article on West Kootenay Power & Light and another on the War Eagle hoist. The latter piece described the hoist as "remarkable . . . the largest hoist ever built to be operated by electricity. It supplies the first link in a new era of power developed in the Kootenay District."[23] The annual report of the Minister of Mines also carried several pages on West Kootenay Power & Light, explaining that the company was "a factor in the mining development of the District, of sufficient importance to merit place in this Report."[24]

Despite the many optimistic pronouncements, however, the new technology was not without its problems. On 19 May 1899, the new electric hoist at the War Eagle mine was bringing five men up the shaft. The skip carrying the miners was nearly at the surface when the hand-operated brake handle on the winch drum worked loose and came off in the operator's hand. This person was standing on a platform and he fell backwards, tumbling to the floor. By the time he had scrambled to his feet and got back on the platform, the skip carrying the men was hurtling down the shaft, out of control. Miraculously, one of the five survived the crash when it hit the bottom. The other four men were killed.

Virtually everyone in Rossland turned out for the mass funeral. The newspaper reported sombrely that the deceased "were not put away in

BCARS B-05123

The Dead Honored

Captain Haas, who is in charge of the local Salvation Army post, . . . gave a short oration on the virtue of the departed comrade. She spoke of the common sorrow that had called the people together and said it was remarkable that not one of the four unfortunate men had a mother, sister, wife or other relative here to mourn for their departure. In their places the people of the city had gathered together to mourn for the loss of these young lives.

— *Rossland Miner*, 24 May 1899

the grave without tears, for during the service there were but few dry eyes among those who were present."

An inquest examined the cause of the four deaths and witnesses were questioned at length as to the hoist's safety and reliability. Perhaps the most telling evidence came from Rock Hull, the hoist operator. He had never worked an electric hoist until coming to the War Eagle about a month earlier. The accident had occurred during the night shift. Hull, who worked thirteen hours straight when he was on nights, had not had time to examine the equipment when he came on shift because he had to take over the controls right away.[25]

Although West Kootenay Power could scarcely be held responsible for the accident, staff became anxious that other mines might shy away from using electricity as a result of the tragedy. As it turned out, business did not suffer. Relations between the company and the War Eagle mine certainly did, however. The two became embroiled in a law suit and West Kootenay subsequently found it very difficult to collect its bills. In his annual report for the period 1899–1900, Campbell reported tersely that "our business relations with this company have not been very successful."[26]

Campbell also found it difficult to deal with another of the company's major customers, the Trail smelter. Originally built by an American "copper king" from Butte, the Canadian Pacific Railway took over the smelter in early 1898. The purchase was part of the CPR's aim to control the rail traffic of the Kootenays. The company had made substantial investments in the region — acquiring the ships of the Columbia and Kootenay Steam Navigation Company as well as building the rail line through the Crowsnest Pass — where it faced stiff competition from an American rail line, the Great Northern.

The CPR assumed control of the Trail smelter in March 1898, putting W. H. Aldridge in charge. To achieve the company's goal of carrying the bulk of the region's ore traffic, Aldridge began to diversify the plant. He shut down the smelter to carry out improvements and expansion, changes that would make it possible for the smelter to treat the silver lead ores of the Slocan and East Kootenay, as well as Rossland's copper gold ores. Although this required capital investment, the capacity to treat silver lead ores would provide the smelter with a far greater range of potential customers. If the smelter could persuade mine owners in the Slocan and the East Kootenays to ship their ores to Trail, the CPR stood to gain both by carrying the ore to the smelter as well as through any profit that the smelter itself might make.[27]

The changes that Aldridge made to the Trail smelter involved greater reliance on electric power. To guarantee electric supply, he approached West Kootenay Power. In the spring of 1898, he and Lorne Campbell negotiated a contract on behalf of their respective companies. Both men were used to driving hard bargains and were clearly a match for each other. When Doull inquired about the progress of the talks, Campbell replied that nothing had been settled and that he was "finding Aldridge a very difficult man to do business with."[28] The two men did eventually sign an agreement and when the re-furbished smelter began operations again, it used West Kootenay power. The relationship between the two companies — or at least between Aldridge and Campbell — remained difficult.

Ten months later — in March 1899 — Aldridge and Campbell were once again at odds. Campbell had billed the smelter $345 for its February power. Aldridge insisted that only $125 was due. Campbell returned the cheque for that amount on 20 March, threatening to disconnect the smelter's power supply at the end of the month if he did not receive full payment. In his letter to Aldridge, Campbell acknowledged that West Kootenay's Montreal directors — presumably Doull and Hosmer — had helped the smelter to secure a favourable contract with West Kootenay. Campbell wanted to make it clear that things were now a little different and that Aldridge ought not to assume he could trifle with Campbell: "at the present time, we have considerable more swing and we do not wish you to imagine for one minute that this letter is by way of a bluff."[29] Presumably the two settled their differences for the smelter continued to use West Kootenay power. In fact, over the next dozen years and more, changes in metallurgical practice meant that an ample electrical supply became essential for successful smelting and refining.

Aldridge's efforts to put the Trail smelter on a solid basis after 1898 encountered a number of obstacles. Strikes in 1899–1900 and 1901–1902, for example, closed many mines. Without an adequate ore supply, the smelter could do very little. In 1901, the Minister of Mines' Annual Report noted Trail's inactivity: "The copper furnaces . . . were only operated to any extent during the first six months of the year. . . . Only one lead furnace was run (there are three) during most of the year."[30] Making a bad situation even worse was the state of the US market. In 1901 the smelter conglomerate, American Smelting and Refining Company [Asarco], established a near monopoly in the ore treatment industry south of the border. It became next to impossible for Canadian suppliers to sell lead in the United States.[31]

With the lucrative US lead market in the hands of Asarco, BC's silver lead mines had nowhere to ship their product for final treatment. The province's mine owners responded with "a very active agitation for Government assistance in the development of lead mining and smelting."[32] During the spring of 1901, public meetings were held throughout the Kootenays. The outcome of these gatherings was that "on April 15 a large deputation from the Kootenay district, generally, waited upon the members of the Government at Ottawa with the object of securing a bounty towards establishing the lead refining industry in Canada."[33] Within six weeks the federal government complied and enacted the called-for bounty (evidently the first ever given in Canada).[34] The legislation, to take effect in 1902, involved an initial payment of five dollars for each ton of lead produced at a Canadian refinery, with the proviso that the lead had also to have been mined and smelted in Canada. For its part, the CPR agreed to build a lead refinery at Trail, thereby guaranteeing that BC's lead ore could be treated reasonably close to home.[35] Naturally, both the lead ore and the final product would be transported on the CPR tracks.

The CPR had considered selling the smelter but with the lead bounty in place, the company renewed its commitment to Trail's expansion.[36] In 1902 it built the promised lead refinery, taking advantage of new technology. The refinery used the Betts process of electrolytic refining, the process's first ever commercial application.[37] At first, the refinery's capacity was a modest eight tons a day but its success was such that this had doubled by 1904. Two years later the plant was handling seventy tons a day.[38]

These developments were critical ones not just for Trail but also for West Kootenay Power. As the smelter grew and relied more and more on metallurgical processes requiring electric power, West Kootenay assumed a critical role in its operations. The outcome was, perhaps, predictable. In February 1905, Thomas Shaughnessy, president of the CPR, wrote to advise his Vancouver superintendent that "we have purchased control of the West Kootenay Company," adding that "it will be of advantage as a factor in keeping the mining and smelting interests of the Kootenay country in close touch with us. It is intended to expend nearly a million dollars in increasing the power."[39] Ten years later, the Consolidated Mining and Smelting Company of Canada [Cominco], itself a CPR subsidiary, formally purchased West Kootenay Power & Light.

The Rossland mines and the Trail smelter were not the only industrial customers to whom West Kootenay hoped to supply power. When an electric street railway was planned for Nelson, the company quickly built a power line into the city. In his report for 1899, Lorne Campbell assured his directors that "The Nelson extension is well underway, and we fully expect to have this completed not later than the 15th November, and we will immediately start in and supply power to the Nelson Tramway Co."[40] By his calculations, this would yield some $800 a month in revenue. At the same time, Campbell offered an optimistic assessment of the company's future. All in all, he assured the directors, "business prospects are exceedingly bright."

Campbell's confidence seemed well-founded. The company had been producing power for little over a year and both demand and revenues were climbing. Not only was the company about to tap the business of Nelson, it had also recently contracted to purchase the pioneer Rossland Water and Light plant. As the last chapter suggested, its owners had made little effort to expand either part of their business. Once West Kootenay power came into Rossland, the company stopped trying to produce its own power and simply purchased its electricity from West Kootenay. No alternative suppliers of water were on the scene, however. Municipal authorities became increasingly irritated and finally decided to purchase the water business of the company, giving West Kootenay the opportunity to buy the electrical division. Campbell welcomed this, describing the purchase as "one of the best deals which the WKP&L have entered into, and one which I am safe in saying they will never regret." Supplying domestic power, he pointed out, was a good deal more profitable than industrial power and guaranteed the company a handsome income. Expansion was going ahead on other fronts as well. A third generator was being installed at Lower Bonnington, raising generating capacity by two thousand horse power. The original two machines each produced approximately a thousand horse power, so when the new machine came on line in early 1900 the plant at Lower Bonnington was producing four thousand horse power.

The company paid for the costs of expansion by issuing more stock. At a meeting in the autumn of 1899, the original capital of $1,100,000

> ### *The Need to Grow*
>
> The general situation of affairs at the present time is in your own hands to decide whether you will go ahead and have a large plant here or whether you will be content to operate our present plant and take everything there is out of it. . . .
>
> Our net revenue warrants a further development of power and I trust you will be able to give me authority to further extend our plant and increase our revenue.[41]
>
> — **LORNE CAMPBELL** addressing the West Kootenay Power's directors in 1900

was raised to $2,100,000. Two Montreal men received a million dollars in West Kootenay shares, in exchange for meeting the expenses of the various projects underway. At the next annual meeting—in the autumn of 1900—Campbell urged the directors to continue to expand generating capacity.

Campbell felt that running all three generators continuously at full capacity was placing too great a strain on them. In any event, an accident or breakdown was bound to occur sooner or later and then customers would be inconvenienced. He recommended building a new plant with a capacity of ten thousand horse power. The directors were more cautious than their general manager. They arranged for an outside engineering report.

The 1901 report—carried out by J. G. White & Company of New York—provided a lengthy assessment of the position and prospects of West Kootenay Power & Light. It concluded that while the company was in excellent financial shape, "the enterprise lacks the stability of an electric company serving a large city or an established manufacturing community, and in general partakes to a considerable extent of the general risk which attends the mining operations of the West Kootenay District."[42] As this passage suggests, the author of the report felt that mining camps were not a reliable base on which to maintain an electric utility. To compensate for mining's ups and downs, he recommended that the company supply power to as many camps as possible. In particular, the report emphasised the need to move quickly and provide power to the newly-established copper mines and smelters of the Boundary district directly west of the Kootenays. On the basis of the report, West Kootenay Power opted to expand.

NOTES

[1] See "Appendices: Petitions for Private Bills, Etc.," *Journals of the Legislative Assembly of the Province of British Columbia*, Vol. XXVI, 1897, i-lxi, and "Charters For Water Power," *Rossland Miner*, 2 March 1897, 2.

[2] See the discussion of this group's role in Rossland's development in Mouat, *Roaring Days: Rossland's Mines and the History of British Columbia*, Vancouver: UBC Press, 1995, 17-22.

[3] Details of Largey's career are from "P. A. Largey Murdered," *Rossland Miner*, 12 January 1898, 1; "The Killing of Largey," *Rossland Miner*, 13 January 1898, 1.

[4] "Early Days in Rossland," *Northwest Mining Truth* (Spokane), 2 July 1921, 8-10, an article that details the reminiscences of Durant. See also "Rossland and Its Builders— Oliver Durant, Esq.," *The Mining Record*, Vol. II, No. 9 (September 1896): 25.

[5] *Canadian Men and Women of the Time*, Toronto, 1912, 548. In addition to this source, detail on Hosmer's career may be found in *The Canadian Who's Who*, 1910, 109-10; Charles G. D. Roberts & Arthur L. Tunnell, eds., *A Standard Dictionary of Canadian Biography: The Canadian Who Was Who*, Toronto, 1934, Vol. 1, 270-71; "Charles R. Hosmer, Financier: Who He Is and What He Does," *Saturday Night*, Vol. 23, #39 (9 July 1910), 19; and "Canadian Pacific Railway Company's Long Copper Wire," *Canadian Electrical News and Steam Engineering Journal*, Vol. 8, No. 6 (June 1898), 113-14 (the article includes a brief biography as well as a photograph of Hosmer). For a valuable discussion of the early years of the telegraph in Canada, in which Hosmer was heavily involved, see the first chapter of Lawrence Surtees' *Wire Wars: The Canadian Fight for Competition in Telecommunications*, Scarborough: Prentice-Hall Canada, 1994.

[6] The first quotation in the paragraph is from "Petition," *Journals of the Legislative Assembly of the Province of British Columbia*, Vol. XXVI, 1897, v; the second from "Petition," *ibid.*, xvii. Sir Charles Ross, William Doull, Herbert Holt, and Frank Paul were the signatories of the second petition, which sought to create the "Kootenay Power & Light Company." However, the *Rossland Miner* assured its readers that "The promoters of this company are really the same as the promoters of the West Kootenay company" ("Charters For Water Power," *Rossland Miner*, 2 March 1897, 2).

[7] Detail here is largely from his obituary, "Sudden Death of Mr. Wm. M. Doull," *The Gazette* (Montreal), 6 October 1913, 4, and the quotation is from this source. It is clear from his career that Doull figured in the business group described by Armstrong and Nelles in their book *Southern Exposure: Canadian Promoters in Latin America and the Caribbean 1896–1930*, Toronto: University of Toronto Press, 1988, although their book does not refer to him by name.

[8] In a telegram dated 25 May 1896, Hosmer stated that Sir Charles Ross had left Montreal the day before and would like an engine pass (the telegram survives in the Van Horne Papers, RG1, #80968, Canadian Pacific Archives, Windsor Station, Montreal). Engine passes permitted the traveller to go up and sit on the locomotive as it sped along the tracks, at the time a popular thrill among railway passengers.

[9] Roger Phillips, *A Historical Appraisal of the Ross Rifle*, Antigonish, NS, 1984, 15. The first chapter of this book provides a very useful summary of Ross's unusual life; see also his obituary: "Ross Rifle Inventor Dies," *Vancouver Daily Province*, 29 June 1942, 8, and *The Times* (London), 30 June 1942, as well as the entry in *Who's Who, 1911*, 1750. His divorce case was covered in various issues of *The Times*: "The Ross Divorce Case," 20 May 1897, 12; 22 May 1897, 14; 24 May 1897, 8; and 14 June 1897, 8. At least one of

these stories was reprinted in a BC paper: "The Ross Divorce" (subtitled, "Airing of the Troubles of the Titled Kootenay Miner"), *Victoria Daily Colonist*, 17 June, 1897, 7, reprinted from *The Times* of 22 May.

10 "Tales By the Town Crier," *Rossland Miner*, 3 February 1909.

11 "Its Great Wealth," *Montreal Gazette*, 27 November 1897, 6; see also "From Western Contemporaries," *The British Columbia Review*, 4 September 1897, 200. The original agreement between Ross and the West Kootenay syndicate survives in the Ross Papers, held by the National Archives in Ottawa (Manuscript Group 30, A95, Volume 1, file IIA). A photocopy is in the History files of the Cominco Papers in Victoria, at the BC Archives and Records Service (Box 427, File 1, Add. Mss. 2500).

12 See for example the letters of Lorne Campbell (then of Canadian General Electric's Engineering Department in Toronto) to Ross, 16 September and 28 October 1897. The first of these refers to "your rough sketch which accompanied your letter of August 21st"; in the second, Campbell assured Ross that "We have instructed Schenectady [General Electric's US headquarters] to prepare the exciter for the reception of leather link couplings same as suggested in your letter."

13 "Early Days in Rossland," *Northwest Mining Truth*, 2 July 1921, 8.

14 On this episode, see esp. Roger Phillips, *A Historical Appraisal of the Ross Rifle*; A. M. Willms, "Decision Making: The Case of the Ross Rifle," *Canadian Public Administration*, Vol. II, No. 4 (December 1959): 202-13; and Ronald Haycock, "Early Canadian Weapons Acquisition: '—That Damned Ross Rifle,'" *Canadian Defence Quarterly*, Vol. 14, No. 3 (Winter, 1984/85): 48-57.

15 Detail in this paragraph is drawn chiefly from the first two chapters of Roger Phillips' *A Historical Appraisal of the Ross Rifle*.

16 Lorne A. Campbell to Wm. M. Doull, 1 June 1898, copy in West Kootenay Power & Light historical notes, "1898," file 3, held at South Slocan archives. Biographical detail here is from the vertical file on Lorne Campbell (microfilm reel 22, exposures 2495–2513), at the BC Archives and Records Service, Victoria.

17 The Gooderham-Blackstock syndicate's purchase of the Centre Star mine for $2,000,000 was reported in "Rossland Notes," *The Mining Record*, Vol. 4, No. 9 (September 1898): 39-40. This was the same group that had purchased the War Eagle mine—neighbour to the Centre Star—in December 1896.

18 Durant is not listed among the shareholders at the annual general meeting held in Rossland 17 October 1900 (photocopy in West Kootenay Power & Light historical notes, "1900," file 5, held at South Slocan archives). The minutes of the 1899 general meeting "acknowledged with regret Oliver Durant's continued illness and withdrawal from active participation in company matters."

19 For an excellent discussion of the public perception of electricity during this period, see the chapter, "Hydro as Myth," in H. V. Nelles, *The Politics of Development: Forests, Mines and Hydro-electric Power in Ontario, 1849–1941*, Toronto: Macmillan of Canada, 1974, 215-55. The chapter includes a discussion of the way in which Niagara's hydroelectric power stimulated up-state New York.

20 Geo. P. Low, "The Kootenay-Rossland Power Transmission," *The Journal of Electricity*, Vol. VII, No. 5 (May 1899), 70. Cf. "Developing an Immense Water Power," *Canadian Electrical News and Steam Engineering Journal*, Vol. 7, No. 7 (July 1897), 130-31; "Progress of the Kootenay Power Plant," *Canadian Electrical News and Steam Engineering Journal*,

Vol. 8, No. 2 (February 1898), 21-22; "Long Distance Transmission Plant of the West Kootenay Power and Light Company," *Canadian Electrical News and Steam Engineering Journal*, Vol. 8, No. 9 (September 1898), 173-76.

21 Geo. P. Low, "The Kootenay-Rossland Power Transmission," *The Journal of Electricity*, Vol. VII, No. 6 (June 1899), 87.

22 *Ibid.*, 88.

23 "The War Eagle's New Electric Hoisting Plant," *The Mining Record*, Vol. V, No. 1 (January 1899), 26; see also Horace F. Evans, "West Kootenay Power and Light Co.," *The Mining Record*, Vol. V, No. 7 (July 1899): 25-28; "Electrical Apparatus for Mining," *Canadian Electrical News and Steam Engineering Journal*, Vol. 8, No. 5 (May 1898), 84; and "Electrical Apparatus for the War Eagle Mine," *ibid.*, 85.

24 "West Kootenay Power and Light Company," British Columbia, *Sessional Papers*, 1899, 1089-90.

25 "A Strict Inquiry," *Rossland Miner*, 23 May 1899, 1, 6. The transcript of the inquest—103 pages of testimony—survives in the BC Archives: British Columbia, Attorney General, Inquisitions, 1872–1937, GR1327 (microfilm reel B2377, 76/1899).

26 Quoted in West Kootenay Power & Light historical notes, "1900," file 5, held at South Slocan archives.

27 The CPR had several connections between Trail and the Slocan mines: it controlled the Nakusp and Slocan Railway which linked Sandon to the Upper Arrow Lake where CPR steamers operated, and the company also owned branch lines which connected Slocan City with Trail. The Moyie area's silver lead ore could be brought to Trail on the CPR's Crowsnest Pass railway and its Kootenay Lake steamers. For details of these connections, see George A. Tripp, "Transportation and Lead Smelters in the Kootenays; a Reconsideration." BA essay, University of British Columbia, 1970, and Cole Harris, "Moving Amid the Mountains, 1870–1930," *BC Studies*, 58 (Summer, 1983): 3-39.

28 Lorne A. Campbell to Wm. M. Doull, 1 June 1898, copy in West Kootenay Power & Light historical notes, "1898," file 3, held at South Slocan archives.

29 Lorne A. Campbell to W. H. Aldridge, 20 March 1899, copy in West Kootenay Power & Light historical notes, "1899," file 4, held at South Slocan archives.

30 British Columbia, *Sessional Papers*, 1902, 1049.

31 For the rise of Asarco, see James E. Fell, Jr., *Ores to Metals: The Rocky Mountain Smelting Industry*, Lincoln, 1979, 221-54, and Isaac F. Marcosson, *Metal Magic: The Story of the American Smelting & Refining Company*, New York, 1949. Christopher Schmitz's "The Rise of Big Business in the World Copper Industry 1870–1930" (*Economic History Review*, 2nd Series, Vol. XXXIX, No. 3 [1986]: 392-410) provides an overview of the consolidations within the mining and metallurgical industries.

32 *Morang's Annual Register of Canadian Affairs*, 1901, 55. Years later, Aldridge claimed that he was responsible for the bounty: see the typescript interview of Aldridge by Howard Bayley (Cominco's Supervisor of Publicity), May 1954, 5 (a copy of this interview is held in the Cominco file of the Rossland Historical Museum's Archives).

33 *Morang's Annual Register of Canadian Affairs*, 1901, 56. See also *The Mining Record*, Vol. VIII, No. 6 (June 1901), 175, and T. G. Blackstock to Laurier, 11 April, 1901, Toronto, #55277, MG26, G, Vol. 193, Laurier Papers, National Archives of Canada.

[34] CANADA *Debates*, 1901, Vol. II (21 May 1901), 5716-29. The previous Conservative government had also provided a bonus system to encourage the BC's lead industry, offering incentives for smelting silver-lead ores (CANADA *Debates*, 1895, Vol. II [5 and 18 July 1895], 3925-28, 4763-67). O. Mary Hill claims that this was the first Canadian bounty, in her book, *Canada's Salesman to the World: The Department of Trade and Commerce, 1892–1939*, Montreal, 1977, 32-33.

[35] Shaughnessy (President of the CPR) to Sir Richard Cartwright (federal minister of Trade and Commerce), 24 December 1901, MG28, III, 20, microfilm reel M-3042, Shaughnessy Letterbooks, CPR Records, National Archives of Canada.

[36] Rumours of the smelter's sale to the Gooderham/Blackstock group (owners of both the War Eagle and Centre Star mines in Rossland) were occasionally reported. See, for example, *Grand Forks Miner*, 19 August 1899, and *The Mining Record*, Vol. VIII, No. 5 (May 1901), 169-70. As late as 1903, however, Aldridge was writing to Shaughnessy on the "Advantages of Trail Smelter to Railway" (Aldridge to Shaughnessy, 6 December 1903, Cominco Historical Files, Microfilm #8, Add. Mss. 2500, BC Archives and Records Service).

[37] The process is explained by its inventor, Anson G. Betts, in "Electrolytic Lead-Refining," *Transactions of the American Institute of Mining Engineers*, Vol. XXXIV, (1904): 175-84; and A. G. Wolf, "The Betts Process at Trail, BC," *Mines and Minerals*, August, 1907, 11-15. Both smelter and refinery processes are described by A. J. McNab, "Lead Smelting and Refining Practice at Trail, BC," *Journal of the Canadian Mining Institute*, Vol. XII (1909): 424-36, and in a form more intelligible to the non-expert, with pictures, by J. M. Turnbull, in "Trail Smelter and Lead Refinery," *Mines and Minerals*, September, 1910, 121-25. Turnbull, McNab and Betts all worked for the Canadian Smelting Works, although Betts' tenure was brief. He supervised the installation and early operation of the refinery.

[38] Wolf, "The Betts Process," 11; McNab, "Lead Smelting and Refining Practice at Trail," 424.

[39] T. G. Shaughnessy to R. Marpole, 21 February 1905, Canadian Pacific Railway, Shaughnessy Letterbook 86, December 1904–June 1905, MG28, III 20, National Archives of Canada.

[40] Quoted in West Kootenay Power & Light historical notes, "1899," file 4, held at South Slocan archives.

[41] [41] Quoted in West Kootenay Power & Light historical notes, "1900," file 5, held at South Slocan archives.

[42] J. G. White & Company, "Report on West Kootenay Power & Light Company, Limited," typescript dated 20 July 1901, 7 (copy held in West Kootenay Power office, Trail). This report comprises ninety-one pages of text, as well as an additional section of photographs and a fold-out map. The first eight pages are in the form of a letter to the Anglo-American Debenture Corporation, Ltd., 75 Lombard Street, London.

CHAPTER THREE

The Company Triumphant: *Into Nelson and the Boundary District*

In the first years of this century, West Kootenay Power expanded well beyond the communities of Rossland and Trail. The company's growth — achieved by increasing its generating capability, extending its lines, and securing new business — might seem straightforward enough, yet the sale of power in nearby Nelson, as well as to mining and smelting interests in the Boundary mining district, led to confrontation and conflict. In part, these disputes reflected a competitive business environment, but other factors were also at work.

In Nelson, the company encountered a formidable political opponent, newspaper editor and some-time mayor John Houston. Houston regarded West Kootenay Power as a grasping monopoly that would squeeze the last ounce of profit from consumers. Arguing that the city would be better off running its own utilities, Houston persuaded Nelson residents to build a competing power facility of their own on the Kootenay River. West Kootenay Power's General Manager, Lorne Campbell, was fiercely loyal to his employer and viewed Houston as a bitter personal enemy. Sparks flew whenever the two men challenged each other.

In the Boundary district, personalities were not the issue, but trouble still accompanied expansion. Business rivalry — between West Kootenay Power and another utility, as well as between the Canadian Pacific Railway and the competing American line, the Great Northern — was fierce. As a result, West Kootenay Power encountered numerous obstacles when it tried to build transmission lines into the new mining district.

West Kootenay Power's difficulties are a reminder that the company never operated in a vacuum. To make sense of its development, we need to look at the bigger picture, the context in which it operated; only then can we understand the problems that it faced in Nelson and the Boundary district.

The 1901 report on West Kootenay Power advocated expansion, noting the precariousness of the resource economy upon which the company relied. The boom and bust cycle of the mining industry was a risky basis for any business. Lorne Campbell was aware of the need to diversify and knew that domestic supply was a market that the company could not afford to overlook. The move into Nelson was clearly a step intended to achieve an important goal as well as evidence that the company was determined to grow.

West Kootenay Power's transmission lines into Nelson were not the first to bring electricity to that city. Nelson had enjoyed the benefits of electric light for some time. In fact, the province's first hydro-electric plant began operations there in early 1896 and plans for its construction had begun four years before.[1] In 1892, a group of Nelson business people received a provincial charter for a power company. Most of them had little or no knowledge of electrical engineering but all had confidence in Nelson's future.

Unlike West Kootenay Power, the Nelson Electric Light Company was not a sophisticated enterprise nor were its backers as well-connected. The company's modest size and financial resources help to explain why its early years were difficult ones. Unable to afford a skilled engineering staff, the company had more than its share of embarrassing breakdowns and accidents. It also had some bad luck. Soon after the company's formation in 1892, the price of silver plunged and depression gripped the mining industry. Then the company discovered that its operations would encroach upon the property of a well-connected resident who was less than enthusiastic about their project. And the site itself was not ideal: the amount of water in Cottonwood Creek was scarcely adequate for reliable power generation.

The plant on Cottonwood Creek began generating electricity in February 1896. Its capacity was modest and Nelson consumers soon discovered that their electrical supply was often erratic. They also had to deal with an erratic business person, one John Houston, the key figure behind the Nelson Electric Light Company and probably the dominant personality in Nelson. It is hard to define Houston's role in the life of Nelson: he was an entrepreneur, a politician, a newspaper editor. A well-known and popular figure, Houston was elected as Nelson's first mayor in 1897 and acclaimed to the position the following year. He was hard-drinking, argumentative, and inconsistent. He would periodically drop everything and disappear from town for months at a time. Even a friendly account of his life chided Houston for possessing "the most inexplicable idiosyncracies of any man that ever did as he blamed pleased" and his own newspaper confessed that "his idiosyncracies are so numerous and marked that he might do almost anything."[2] Houston was thirty-nine years old when he came to Nelson in 1890. Since his boyhood on an Ontario farm, much of his life had been spent wandering western North America. An experienced newspaper man, Houston established Nelson's first paper shortly after his arrival that spring of 1890. He soon became involved in real estate as well as several other ventures, including the Nelson Electric Light Company.

The 1890s were boom years in the Kootenays and Nelson grew along with many other communities in the region. The town's central location, and its growing importance as a railway and shipping hub, helped it to become the regional centre. And people believed that the future would bring great things. The comments of a CPR executive in 1898, reprinted in a publicity-seeking brochure entitled *The City of Nelson: The Metropolis of the Far-famed Mineral District of Kootenay*, were typical: "We have made Nelson our central point for the interior of this Province because we feel sure that this city is destined to become the Birmingham of British Columbia."[3] Although Nelson never became the manufacturing centre of the province, this was not for lack of trying. Houston and others were tireless boosters. The Nelson Electric Light Company was intended to help the city's growth as well as to return a profit to its backers.

When the privately-owned power company failed to live up to expectations, Mayor Houston decided that the newly-incorporated city ought to purchase the company. Persuading Nelson's residents to endorse this move was not easy and the 1898 referendum on the issue was highly controversial. When the ballots were first counted, a slender majority rejected the purchase by the city. A re-count, however, reversed the decision. (Years later one of the officials allegedly confessed to tampering with the ballots prior to the second count.[4]) A court case followed, challenging the legality of the city's purchase. The dispute ended up in the BC Supreme Court where the judgment found that the acquisition was indeed lawful, despite the fact that the mayor and several aldermen had close links to the private company.[5] And so Nelson acquired its own electrical company.

Municipal ownership was not unusual around the turn of the century. The cities of Edmonton and Winnipeg both acquired privately-owned electrical plants and under circumstances not unlike those in

Looking every inch the hard-living newspaperman, John Houston poses with an unnamed boy.

KOOTENAY MUSEUM ASSOCIATION AND HISTORICAL SOCIETY, 1174

BCARS B-04877

If the amount of overhead wiring is any indication, Nelson lived up to its reputation as the Electrical City. The approaching tram does not appear to be causing much excitement on Baker Street.

Nelson.[6] The idea behind such purchases was to hasten the growth of the city. A municipally-owned plant could attract industry by offering cheap power. As the mayor of Medicine Hat explained, "Municipal ownership and industrial progress go hand in hand."[7]

An Experiment In the Municipal Ownership of Electric Power . . .

Along the lines of municipal ownership of public utilities the city of Nelson, British Columbia, is ahead of Glasgow, Birmingham, Toronto, or any of those corporations that are held up as model municipalities, and it is at the present the theatre of an industrial experiment decidedly more daring than anything any city in the world has so far attempted.

. . . The first use of this power [of the Kootenay River] was made by the West Kootenay Power and Light company, organized . . . by Sir Charles Ross. This company supplies power to the cities, mines and smelters and has been a profitable venture. The man who developed this rich and powerful corporation was an aristocrat and speculator and is now a wealthy capitalist. About the same time another corporation was being developed by another man who combines the functions of newspaper editor, mayor and member of the legislature. That corporation is the city of Nelson. So far as the community of Nelson is concerned the interests of these two corporations were antagonistic. The struggle between them has been very interesting and at times exciting. The city has held its own and its citizens are deriving great present and prospective advantages from the successful fight. . . .

— D. B. BOGLE, "An Experiment In the Municipal Ownership of Electric Power Being Carried Out at Nelson, BC," *Manitoba Free Press*, 1 July 1905, 27

The purchase of the Nelson Electric Light Company meant that West Kootenay Power encountered a competitor when it decided to bring power into the city—and it would have to deal with John Houston. Houston suffered politically as a result of the controversy in 1898 over the sale of the power company to the city: he was voted out of office in the January 1899 election. He was not out for long, however: the next year he was mayor again and also elected as the member for the provincial legislature. During the year he was out of office, West Kootenay Power signed a deal to supply power to Nelson's new street railway.

Just before Christmas 1899, Lorne Campbell proudly drove the first street railway car down Nelson's streets. The company that built the system — the Nelson Electric Tramway Company — was a subsidiary of a large British enterprise, the British Electric Traction Company.[8] (The latter business was likely persuaded to make the move into Nelson by shareholders who were also interested in a local mine.) The agreement between West Kootenay Power and the Nelson Electric Tramway Company stipulated that the street railway would have to buy a minimum amount of power. This clause allowed West Kootenay Power a guaranteed return on the initial investment of power lines into Nelson. For the street railway, the provision meant that it had either to run a substantial number of cars or to sell off the surplus power. With a population of just over five thousand in 1901, Nelson was not a large city. Attracting enough passengers from this slender base to run cars that would consume all the power that street railway had contracted to buy was going to be a difficult task. The company probably knew from the beginning that the second option — selling off the excess power to others — was the more likely scenario. Houston opposed such sales, arguing that they would reduce the market share of the municipally-owned plant. He was determined to fight any move that would weaken the viability of the City plant.

Houston was back in the mayor's chair in 1900 and thus well-placed to fight West Kootenay Power. He began by attempting to improve the water supply to the struggling plant on Cottonwood Creek. The plant itself had been re-furbished in 1899 and now Houston decided to raise the height of the dam by four feet. While the rationale made sense, the results were disastrous: the dam collapsed shortly after the work was finished in the summer of 1900. A string of equally embarrassing incidents over the next two years — the creek froze and brought operations to a halt, another washout crippled the water supply, then the generator burned out (twice) — must have left city residents disillusioned with the virtues of municipal ownership. Houston, in the midst of all this, decided to turn his attention to provincial politics and left the affairs of the city to another, although one of his last gestures was to take out a water license on behalf of the city on the Upper Bonnington Falls of the Kootenay River. He may have moved on but the battles were not over and he would be back.

As a result of the problems with the Cottonwood plant, the city of Nelson was forced to buy electricity from West Kootenay Power. With Houston once more out of office, the company finally achieved its goal of supplying power to various businesses in the city, notably Pat Burns' meat-packing plant and the Hall Mines smelter. However, its original customer, the tramway company, ran into financial difficulties. Its ridership was not high enough to earn a profit on the money invested. In the

BCARS B-05120

Construction work in 1906. Above, workers pose amid the bedrock of West Kootenay Power's Upper Bonnington power site. Below, erecting poles for the transmission line into the Boundary district. Lorne Campbell is there to keep an eye on operations, on horseback on the far left.

ROSSLAND HISTORICAL MUSEUM, 597-15

autumn of 1904 the operation of Nelson's street railway was handed over to the city. (Five years later, after a disastrous fire brought an end to the struggling business, the city purchased it outright.)

Meanwhile, Houston's career as a provincial politician was on the slide. Not surprisingly, he was something of a maverick in the legislature. He was in the news in the spring of 1902 after he fought with the Speaker of the House. Accused of being drunk at the time, he denied the charge: "Mr. Houston said that he was not drunk then, he had had four or five big jorums of Scotch whiskey, but was sober." And in any event, as he explained to his Nelson constituents, "if every member who got drunk were excluded from the House, they would not often have a quorum."[9] But Houston paid a price for his behaviour. As head of the province's Conservative Association, he could well have benefitted from the electoral success of Richard McBride in 1903. McBride, himself a Conservative and the province's first premier to take office with an organised party behind him, might in other circumstances have appointed Houston to the cabinet. But this McBride decided not to do. When Houston discovered that he had been passed over and would not become a member of the cabinet, he levelled angry charges against both the Lieutenant Governor and Premier McBride. He made the headlines for a few weeks but his chances of a significant political career were over. He decided to return to his Nelson stronghold and in 1905 Houston was once again acclaimed mayor of the city.[10]

Back in the mayor's chair, Houston finally accepted the inevitable: the Cottonwood plant would never become a viable operation, and he swung his support behind an alternative plan. Houston recommended that the city borrow $150,000 in order to build a large generating plant on the Kootenay River, upstream from West Kootenay Power's operations at the Lower Bonnington Falls. The idea had been in the works for several years. In 1904 the city of Nelson had acquired a second water license on the Kootenay River, supplementing the one that Houston had obtained in January 1901. Voters endorsed Houston's proposal for raising money and plans for the new power plant were soon underway.

Lorne Campbell did not welcome the prospect of a competing electrical plant on the Kootenay River. For one thing, West Kootenay Power was in the midst of building a new plant of its own, on the north side of the river directly across from the city of Nelson's proposed site. And in any event, if the city did build a serviceable plant it might be able to do without any West Kootenay power at all.

Campbell had known for some time that the city of Nelson wanted to build a plant at the Upper Bonnington Falls and he had gone out of his way to prevent this from happening. In early 1902 he filed a mineral claim for the only ground suitable for a power plant on the south side of the Upper Falls, that is, where the city held its water right. Even earlier he had claimed the property for a mill site for mining purposes. Naturally enough the city was after the same piece of land and in 1902–1903 the two fought a legal battle over who could legitimately claim it.[11] Campbell was bitter when the city emerged as the victor. He blamed West Kootenay Power's lawyer for botching the case and suspected that Houston had exercised his political influence to secure the ground. When the city tried to raise $150,000 in bonds in 1903, to cover the cost of its proposed plant, Campbell did everything he could to scare away investors. Whether through Campbell's efforts or for some other reason, the city was unable to find the necessary money. Then in 1905 Houston returned to Nelson and took up the cause.[12]

Lorne Campbell Explains . . .

Everything was running along smoothly in connection with my application to purchase this millsite, when Mr. Chester Glass, of Spokane, appeared on the scene with instructions to handle this affair . . . you can thank Mr. Chester Glass, of Spokane, for making a mess of this. . . .

I have not the slightest doubt in saying that the Chief Commissioner of Lands & Works would have dealt fairly in connection with this . . . but Houston, as soon as the hearing was over in Nelson, immediately proceeded to Vancouver and simply held them up for a Crown Grant of the land which they applied to purchase.

You are no doubt aware of the rotten conditions of affairs in general at Victoria, and I am of the opinion that they would sell their bodies and souls or give any local member anything he would ask, providing he would pledge his vote to vote with the Government; and I am of the opinion that this is what Houston did, and is the reason that the City of Nelson have a Crown Grant of this property to-day.

— **LORNE CAMPBELL** to **W. M. DOULL**, 5 February 1903

In the spring of 1905, Houston managed to raise the $150,000 needed to build the city's power plant. Construction was soon underway. That stretch of the river must have been a busy spot since West Kootenay Power crews were hard at work on the opposite bank, building the company's number two plant. Lorne Campbell was still determined to do anything he could to prevent the city from completing its power station. He resorted to the courts once more and in May 1905 managed to secure an injunction against the city. This alleged that construction of the city's plant could damage West Kootenay Power's existing works lower down the river as well as having an impact on the viability of the company's new facility. During the summer of 1905 both parties were in court, West Kootenay Power seeking to have its injunction upheld—which stopped the city's crews from dumping rock in the river—and the city seeking damages against the company for the injunction. On 8 August, Justice Irving rendered his judgment in favour of West Kootenay Power.[13]

It was a bitter blow for Houston. He disappeared from sight for a month and a half, eventually turning up down in Nevada. He wrote back to Nelson, submitting his resignation as mayor. Although he did return to British Columbia, his Nelson days were over.

John Houston Resigns

I should have dropped a line before now but as I was a wanderer I thought it best to defer writing until I had a postoffice address. . . .

I might write something personal, but the least said the better. The decision of judge Irving made me disgusted; that decision coupled with the fact that I seemed to be unable to make any headway in business in Nelson, developed a fit of despondency. . . .

I am discredited. I am broke in my old age; but I gave Nelson and its people the best that was in me for 15 years. . . .

— **HOUSTON**, Goldfield, Nevada, 9 October 1905

Despite Houston's departure, the city of Nelson refused to give up and appealed Irving's judgment to the BC Supreme Court. In the spring of 1906 the Chief Justice allowed the city's appeal and quashed West Kootenay Power's injunction. His judgment offered a sarcastic assessment of West Kootenay Power's attempts to prevent the city from building a power plant on the Kootenay River, making it clear where his sympathies lay. Campbell and the West Kootenay Power's lawyer lost no time in laying plans to appeal the judgement to the Privy Council in Britain. Eventually, however, West Kootenay Power reluctantly accepted the presence of a rival on the river.

Early in 1905, the Canadian Pacific Railway purchased a majority share in the company. CPR management regarded the investment as a relatively safe one but the move represented more than this. It also reflected the bitter rivalry between the CPR and its competitor just south of the border, the Great Northern Railway.

City Wins Hands Down

I may say I am not favorably impressed with the raison d'etre of this suit . . . having regard to the fact that the establishment of the defendant's [the City of Nelson's] undertaking means the loss of one of the plaintiff's [West Kootenay Power's] largest customers, and that the plaintiffs were baffled in their attempts, under cover of the mineral act, to occupy the site of the defendant's plant, in order to obtain a monopoly of the power of the Kootenay River. . . .

In my opinion the appeal should be allowed and the action dismissed without prejudice to any future action which may be brought by the plaintiffs.
— **CHIEF JUSTICE HUNTER**, quoted in the *Nelson Daily News*, 7 March 1906

The Great Northern was one of the most successful railways in the United States. The man behind it, Jim Hill, was Canadian-born and had been one of the original group behind the formation of the Canadian Pacific Railway. But he dropped out of that venture and several years later built his own transcontinental railway from his base in St. Paul, Minnesota. Like the CPR, Hill was very interested in securing the business of southern British Columbia. The region lay between the CPR line to the north and the Great Northern to the south. Both railways hoped to carry the ore traffic produced by the booming mining industry and each acquired lines into the Kootenays with that end in sight.

The Canadian Pacific Railway's purchase of the Trail smelter in 1898 was a clear signal that it was determined to profit from the province's mining industry. The company's interest in the Kootenays began not long after it completed Canada's first transcontinental railway in 1885. Four years later, the CPR gained control of the Columbia and Kootenay Railway—a portage line linking the navigable portions of the two river systems—and throughout the 1890s it continued to improve its Kootenay connections. In 1893, the company leased the Nakusp and Slocan Railway, intended to link the Upper Arrow Lake system with the Slocan, completing the line over the next several years. Then in 1896, the CPR opened a spur from Revelstoke to Arrowhead, to improve connections from the Arrow Lakes up to the CPR mainline at Revelstoke as well as to a smelter there. The following year it purchased the Columbia and Kootenay Steam Navigation Company, which ran many of the boats on the lakes and rivers of the area linking the various rail lines. The intent of all this was to facilitate access from the CPR's main line south into the mining regions. However the CPR ultimately concluded that relying upon a spur line down into the

Kootenays from Revelstoke was not going to work. To compete successfully with the American lines, and especially with Hill's Great Northern, the company required more direct access into the region. That was the purpose of the Crowsnest Pass railway, a project first envisaged in 1893 when the CPR acquired Galt's Lethbridge railway in southern Alberta.[14]

In the autumn of 1896, CPR president Van Horne made it clear in newspaper interviews that his company intended to build a line west to the Kootenays through the Crowsnest Pass.[15] The following year the federal government did its part by passing a bill subsidising the CPR's construction of the Crowsnest Pass Railway. Despite critical comments from various quarters — mostly from the coastal communities of British Columbia who argued that the line ought to continue through to the Pacific — Ottawa felt that a rail line through the Crowsnest Pass would result in better access to the Kootenays from central Canada. This in turn meant greater benefit to the country as a whole since it would help to end the export of much of the region's mineral wealth south across the forty-ninth parallel. In 1898, the same year in which the CPR purchased the Trail smelter, the company built its southern line through the Crowsnest Pass to Kootenay Lake.

Hill was also active in 1898, purchasing two lines out of Spokane into Rossland and Nelson respectively. Since his transcontinental passed through eastern Washington and Spokane, the move provided him with excellent access into the region. The mining boom of the early 1890s had already piqued Hill's interest in southeastern British Columbia. For example, in 1892 he considered advancing $200,000 to have a mine developed on the shores of Kootenay Lake and three years later he decided to build a small narrow gauge rail line into the booming Slocan region, to link it with steamers on Kootenay Lake and thence to the Great Northern line at Bonner's Ferry in Idaho. This led to a violent confrontation between Hill's construction crews and those of the CPR, as the two companies raced to complete lines into Sandon, Hill from the east and the CPR from the west.[16]

Naturally, Hill could never expect the same preferential treatment from the Canadian government as that enjoyed by the CPR. Nonetheless, his rail line did enjoy one significant advantage in the contest to secure Kootenay rail traffic: geography. The region's topography — with its valleys running north and south — gave his railway better access to the region. However, the CPR was determined to prevent him from taking business which it felt ought to remain in Canada. The rivalry came to a head in the early 1900s as the two companies sought to secure the ore traffic of the Boundary district.

Prospectors located the Boundary district's vast copper deposits in 1891, the year after the first of the Rossland mines were staked. The low grade ore did not initially attract much attention; people at the time were more interested in speculating in gold properties. However, geologists and canny investors gradually realised that despite their low grade, the Boundary copper deposits were bound to produce a substantial return as long as production costs could be kept to a minimum. The first mines began shipping ore at the turn of the century and their economical methods of mining and smelting soon attracted attention throughout North America.[17] The longevity of Cominco's operations at Trail and Kimberley have overshadowed the significance of mining activity in the Boundary district but in 1905 the Minister of Mines' Annual Report described the Granby Consldiated Mining, Smelting and Power Company — with mines in Phoenix and a smelter in Grand Forks — as "the largest and most important mining and smelting company in British Columbia."[18] Four years later the Annual Report noted that the Granby mines shipped over a million tons of ore in 1909, adding that "cost of mining, including the crushing of the ore to furnace size and delivering same on board cars, was 84 cents a ton — a performance well worthy of special note."[19] During its period of peak production, the Granby mines shipped four thousand tons of ore every day to the company's smelter. Two other mining companies operating in the Boundary were also very productive and owned their own smelters. Such activity created much work for railway companies and inevitably attracted both the CPR and the Great Northern.

56% of the total tonnage of the Province . . .

In the Boundary District the tonnage of ore has increased about 20% over the preceding year, and now amounts to 965,628 tons, being over 56% of the total tonnage of the Province.

This increase is due to the constantly increasing operations of the Granby Co., the other large companies about holding their own. . . .

The costs of mining and smelting have been gradually reduced in this section, thanks to as fine equipments as money could buy, in the hands of intelligent and scientific men, until they are now reported to be about the lowest in the world. To quote from a recent editorial in the leading American scientific journal:

> Ten years ago the idea of smelting for a dollar a ton and mining for $1.10 a ton would have been scouted as impossible. Yet this has been done at the Granby mines. . . .

— Annual Report of the Minister of Mines for 1905

In the summer of 1904, Hill became interested in the leading Boundary company, the Granby Consolidated Mining and Smelting Company. "CPR Is Outwitted," ran the headline in the *Victoria Daily Colonist*, "J. J. Hill Grabs Vast Mining Interests in the Boundary Country."[20] Hill's ability to carry the bulk of the Boundary's freight on the Great Northern irked the CPR. It had pushed rails into the district five years earlier, back in 1899, hoping that the company's Trail smelter would treat all of southeastern British Columbia's ore.[21]

In fact, the CPR's efforts to reap benefits from the mining industry in the Kootenays had not met with much luck. The Trail smelter, in particular, was not flourishing as expected, despite a considerable investment in upgrading the plant. By the early 1900s, Rossland mines—the principal feeders of the smelter—were proving to be less rich than anticipated and, in an effort to rationalise the mines and protect its investment, the CPR pushed to consolidate the various Rossland mining properties with its smelter. The original idea seems to have come from people involved with Rossland's Le Roi mine, a famous property owned by British investors. The CPR's smelter manager, W. H. Aldridge, met one of the Le Roi's directors in the autumn of 1904, later advising the CPR's president that this person was "anxious to make a large consolidation of Rossland and Boundary interests, . . . and I told him that such a consolidation would certainly receive the encouragement of the [Canadian Pacific] Railway so long as they [*sic*] were sure that they would receive the haulage."[22] More than a year passed before anything came of these negotiations. The key players were Aldridge, as the representative of the CPR's interests, and the owners and managers of the Le Roi, War Eagle and Centre Star mines in Rossland. (These last two properties were both owned by the Gooderham/Blackstock liquor interests of Toronto.) As we'll see, in the end the British owners of the Le Roi decided not to join in the consolidation, but the CPR did secure the other two properties, the War Eagle and the Centre Star. And behind it all lurked the threat of the Great Northern, as the newspaper headlines proclaimed: "CPR Frightened By J J Hill Into Buying Rossland Mines."

The person who wrote the *Toronto World* article must have been well-informed: the CPR did merge the two Rossland mines with its smelter, leading in early 1906 to the formation of the Consolidated Mining and Smelting Company of Canada (more commonly known as Cominco).[23]

The Le Roi's reluctance to participate in the merger reflected behind-the-scenes manoeuvring by Hill and his associates. In a confidential letter to a friend, Hill described how he had an agent in Britain purchasing Le Roi shares on his behalf.[24]

CPR Frightened by J. J. Hill . . .

Does the absorption of the Gooderham and Blackstock mining interests in British Columbia by the crowd that has been identified with the CPR smelters at Trail mean that it is a continuation of the fight that has been waged against J. J. Hill . . . in that Province? Those who have closely followed the changes in ownership are of the belief that it [is] but another move in the diplomatic war of commerce.

. . . it would seem that a strategic move was made on the part of the CPR people, and by this they acquired the stock that they needed to preserve their smelting trade and also their haulage to the east.

. . . Under such conditions as have been detailed in the foregoing it does not seem possible that other than amalgamation, or at least the pooling of interests will follow. It has been whispered that there will be a meeting held of the new interests on Monday afternoon.

— *Toronto World*, 26 June 1905

In the winter of 1905, Aldridge went to London in hopes of persuading Le Roi shareholders to participate in the formation of an integrated mining and smelting company. To his chagrin, he discovered that Hill had got there first: "The Great Northern . . . spent large sums of money in purchasing LeRoi stock and perfecting an organization throughout England in collecting proxies."[25] The meeting of Le Roi shareholders which considered his proposal voted it down. But the setback did not phase Aldridge: "We are going right ahead to form our Canadian Company comprising Centre Star, War Eagle, St. Eugene and the Trail

Hill Explains . . .

For some time past, there has been a plan between some Canadian mine owners at Rossland and the Canadian Pacific Railway to consolidate the Rossland mines and the smelter at Trail. To carry this into effect all depends upon their acquisition of the LeRoi mines [sic] and the Northport smelter. By buying LeRoi shares, as outlined in letter attached, we will be able to prevent the consummation of the scheme and hold control of the LeRoi property, which really controls the Rossland camp, and prevent practically the wiping out of the business of our Red Mountain and Northport line which represents the main business of the Spokane Falls & Northern Railway.

— JAMES J. HILL to E. T. NICHOLS, 7 June 1905

Smelter."[26] True to his word, he helped form Cominco several months later. But the whole episode reveals the lengths to which both the CPR and the Great Northern were prepared to go in order to sabotage each other's plans. Which brings us back to West Kootenay Power.

West Kootenay Power's emergence as an important and reliable energy supplier at the turn of the century meant that it could not avoid being drawn into the battle between the two railway companies. At the end of 1904, a year before Aldridge orchestrated the formation of Cominco, he sent a telegram to CPR president Thomas Shaughnessy, advising him of the advantages of acquiring West Kootenay Power: "monopoly of power would have great weight in controlling freight business. . . . West Kootenay Power Co. valuable property for CPR and they should control in preference to Great Northern."[27] The telegram survives in the Canadian Pacific archives, the first item in the file chronicling the company's purchase of West Kootenay Power. The rivalry between the CPR and the Great Northern explains the CPR's acquisition of West Kootenay Power in early 1905.

West Kootenay Power had reasons of its own for attracting the attention of the Canadian Pacific Railway. The 1901 engineering report on West Kootenay Power recommended that the company build transmission lines west into the Boundary district. For this to happen, West Kootenay Power had to make a considerable investment in terms of plant. Not only would the Boundary connection call for considerable labour with the construction of substations, the erection of poles, and the stringing of lines, but also it meant that the company's generating capacity on the Kootenay River would have to be substantially increased. The company would need a healthy injection of capital, something that the Canadian Pacific Railway could readily supply.

In early 1905, CPR management concluded that it ought to proceed with the purchase of West Kootenay Power. When the sale was being negotiated, the railway company agreed to take part in a bond issue which would raise a million dollars for West Kootenay Power.[28] (This suggests one of the advantages enjoyed by West Kootenay Power in its fight with the City of Nelson, who could never hope to match such a sum.) The money from the bond issue was ear-marked for the Boundary extension: it would help pay for the Number Two plant on the Kootenay River as well as the cost of the seventy-eight mile transmission line into the Boundary.

Securing the financial backing of the CPR did not mean that West Kootenay Power's plans to build into the Boundary were assured of success, however, and the scheme soon encountered a major obstacle. The company's original charter, approved by the provincial legislature

> ### *A Done Deal*
>
> Dear Sir Thomas
>
> Will you kindly before leaving for England, confirm our conversation of this morning with a few lines in writing. I stated to you that our Co. had closed contracts with the Granby Co. and with the BC Copper Company. That as a result of the plans being made and prices received for machinery obtained, I was assured that the million dollars would cover all costs of the works contemplated & provide $100,000.00 for present outstanding Bonds. You told me to go right ahead with the work & that your Co. would stand behind me.
>
> — **W. M. DOULL** to
> **SIR THOMAS SHAUGHNESSY**, CPR President, 24 March 1905

in 1897, permitted it to operate within fifty miles of Rossland. The Boundary extension meant that these terms had to be changed and in late 1905 West Kootenay Power submitted a petition seeking such an amendment. But another company—the Cascade Water, Power and Light Company—already operated in the Boundary. It too petitioned the Legislature, asking that the proposed change to West Kootenay Power's charter be refused. The Cascade company pointed out that its Boundary plant represented a substantial investment which was only now beginning to return some interest on the initial capital outlay. And the point of the original charters for both companies were clear: "it was intended that each should have the area assigned to it within which to operate, and in virtue thereof the investments were made. . . . there can be no valid or sufficient reason shown why the said West Kootenay Power & Light Company, Limited, should be allowed to encroach upon its area." [29] The legislature took the objection seriously and referred the bill to its Private Bills Committee.

In February 1906 the four members of the Private Bills Committee spent a week listening to the arguments of each company's lawyer as well as testimony from various witnesses.[30] West Kootenay Power secured the services of D. M. Eberts, the provincial attorney-general, while E. V. Bodwell acted for the Cascade company. West Kootenay Power argued that the Cascade company simply could not provide sufficient power to the Boundary's booming mining industry—the Kettle River did not have enough water to generate a constant and reliable source of power—and it could show that the mining and smelting companies of the region were anxious to see West Kootenay Power come into the region. For its part, the Cascade company stressed

BCARS B-33556

West Kootenay Power's competition. Above, the City of Nelson's ill-fated dam on Cottonwood Creek, eventually abandoned in favour of the contentious site at Upper Bonnington across from West Kootenay Power's Number Two plant. Below, a graphic illustration of the water shortages that plagued the Cascade Water, Power & Light Company in the Boundary district.

ROSSLAND HISTORICAL MUSEUM, 597-19

that it would be bankrupted if West Kootenay Power came into the Boundary as a competitor. It emphasised that it would be happy to come to some arrangement with West Kootenay, and distribute or sell its power in the Boundary. It pointed out that its British shareholders had invested money in the company in good faith and now faced competition that would wreck its business.

E. V. Bodwell Explains . . .

Some years ago, when [the Boundary district] was not at all developed, and there was not business enough in sight to induce the [West] Kootenay Power Company with all their large capital and magnificent plant to enter that District at all, some of the gentlemen whom I represent, and who reside in England and other parts of Great Britain, they conceived the idea of developing water power there to supply the needs of the community. They invested their money in good faith, having received from the Legislature of this Province an Act incorporating them, and giving them a certain area in which their works could be erected and their operations carried on. Acting upon the faith of the Act of the Legislature, and believing that their investment would receive reasonable protection, they spent somewheres in the neighbourhood of $500,000 in building and equipping an electric power plant. . . .

— Submission to the Private Bills Committee, 7 February 1906

This was a sensitive point in British Columbia. Until the First World War, London was the world's financial centre. Damage to one's prestige in that city could make it impossible to raise any further capital. The province's mining industry, in particular, had attracted substantial investment from Britain since the 1890s. Much of this money had been squandered in what were essentially speculative schemes and the province was well aware that its reputation had suffered as a result.

West Kootenay Power presented an able case. It could point to the urgent requests of the leading Boundary mines, seeking assurances that they could obtain cheap and reliable power. For example, Warner Miller—a former New York senator and president of the Dominion Copper Company—appeared as a witness for West Kootenay Power. He was unequivocal: "If we are not able to get electrical power for certain, and continuously every day of the year, all thoughts of future developments to the property will be over . . . what we want is a permanent electrical supply." As far as he was concerned, the Cascade company was stretched to the limit and could not guarantee his company an adequate source of power. Despite such testimony, the objections to the bill remained and the Legislature refused to pass it.

West Kootenay Power now faced the unpleasant possibility of being prevented from selling power in the Boundary. Construction on the Upper Bonnington facility was almost finished and it would soon be generating electricity. Crews had been hard at work for some months, putting the transmission lines into the Boundary and building substations at Greenwood, Phoenix, and Grand Forks. (West Kootenay Power had organized a subsidiary, the British Columbia Construction and Distributing Company, Ltd., to carry out the work in the Boundary.) The failure of the bill forced West Kootenay Power to weigh its alternatives.

In May 1906 Lorne Campbell wrote to a Vancouver lawyer, asking a simple question: "If our Company were to secure the charter of the South Kootenay Water Power Company, dated 8th May, 1897, would this charter enable our Company to do business in Yale County [i.e., the Boundary district] and remove whatever difficulties might spring up in connection with the wording of our West Kootenay charter?"[31] Presumably the reply was affirmative for West Kootenay Power subsequently bought the latter company. The arrangements were a little complex: the South Kootenay company first acquired the assets of the British Columbia Construction and Distributing Company — which included the transmission lines and the Boundary substations — and then merged its operations with those of West Kootenay Power. A special meeting of the South Kootenay Water Power Company approved the manoeuvres in December 1906.[32] West Kootenay Power was finally in business in the Boundary.

The transmission lines began carrying power into the Boundary from West Kootenay Power's Number One plant in the summer of 1906. The following January the new plant on the Upper Bonnington Falls, the Number Two plant, was in operation, providing a huge boost to West Kootenay Power's capacity. The original Number One plant had a capacity of four-thousand horse power after a third machine was added in 1899. By contrast, the Number Two plant began with two eight-thousand horse power units although in later years its capacity was more than doubled.

West Kootenay Power's competitor in the Boundary, the Cascade company, realised that its chances of survival were now slim. In April 1907 the company sold its assets to West Kootenay Power. When it was all over, *The Mining Record* observed that the deal gave "the West Kootenay Power Company complete control of the power business in Yale and Kootenay districts." And it brought to a close, the magazine noted, "the bitterly contested fight between these rival companies a year ago before the provincial legislature."[33]

The extensions into Boundary and Nelson created some bitterness and ill-will. Back in Nelson, at least some city residents still regarded West Kootenay Power as a personal enemy. A brochure produced for the municipal election in 1908, for example, criticised the company in harsh terms and heaped praise on former Mayor Houston and his colleagues.[34]

Look Back Then Look Forward

In 1897, 1898, 1899, and 1900, the people of Nelson had independence and took pride in the policy that gave their little city a name almost national. But that independent spirit began to decline with the advent of the West Kootenay Power & Light Company, in 1901, with its infamous methods of intimidation, bribery, and perjury.

– *The Campaigner*, Nelson, 7 January 1908

Nor was Lorne Campbell prepared to let bygones be bygones. When the City of Nelson was trying to raise money for another bond issue in 1908, Campbell wrote to the company president to let him know, adding "I think it would be well if you could use your influence . . . to see that this bond issue does not go through."[35]

But if expansion had been controversial, it had also made West Kootenay Power an important electrical utility with a national reputation. Its new Bonnington plant was the largest in the province and the twin 60,000 volt transmission lines were at the cutting edge of electrical developments in North America. The future looked bright.

NOTES

1 See Boyd Affleck, "A History of the Cottonwood Falls Hydro-Electric Plant, Nelson, BC," typescript, 1933.

2 The best source on Houston's personality and his career in Nelson is Patrick Wolfe's "Tramp Printer Extraordinary: British Columbia's John "Truth" Houston," *BC Studies*, #40 (Winter 1978-79): 5-31. See also David Scott and Edna H. Hanic's *Nelson: Queen City of the Kootenays*, Vancouver: Mitchell Press, 1972, 49-78. On Houston's final years

in northern BC, see Frank Leonard's *A Thousand Blunders: The Grand Trunk Pacific Railway and Northern British Columbia*, Vancouver: UBC Press, 1995, *passim*. The quotations in the text are reprinted in Wolfe, *op. cit.*, 7.

[3] Quoted in *The City of Nelson: The Metropolis of the Far-famed Mineral District of Kootenay*, Winnipeg: Bulman Bros. & Co., 1899?

[4] Scott and Hanic, *Nelson: Queen City of the Kootenays*, 68.

[5] The judgment is printed in "Corporation Sustained," *Victoria Daily Times*, 20 December 1898, 3.

[6] See John C. Weaver, ""Tomorrow's Metropolis" Revisited: A Critical Assessment of Urban Reform in Canada, 1890-1920," in Gilbert A. Stelter and Alan F. J. Artibise, eds., *The Canadian City: Essays in Urban and Social History*, Ottawa: Carleton University Press, 1984, 464-65.

[7] Quoted in Weaver, "Tomorrow's Metropolis," 465. The equation of industrial progress with the accessibility of cheap power has been an enduring theme in discussions about the shape and future of the province's economy. (Consider, for example, W. A. C. Bennett's advocacy of the Peace River power projects discussed in chapter seven.)

[8] For the early years of the company, see Douglas V. Parker, *Streetcars in the Kootenays: Nelson's Tramways, 1899–1992*, Edmonton: Havelock House, 1992, 20-60.

[9] Both quotations are from "John Houston's Story of Affair," *Victoria Daily Times*, 2 November 1903, 2, which reprints an account of the public meeting held in Nelson on 29 October 1903. For the confrontation between Houston and the Speaker, see "Proceedings of the Legislature," *Victoria Daily Times*, 7 May 1902, 3.

[10] Detail in this paragraph is drawn from Wolfe, "Tramp Printer Extraordinary: British Columbia's John 'Truth' Houston."

[11] The legal proceedings are described in papers of the provincial Dept. of Lands and Works at the BC Archives and Records Service (see GR1132, file 2).

[12] This section is based on comments in "Nelson Mayor is Missing," *Cranbrook Herald*, 21 September 1905, 1, and D. B. Bogle, "An Experiment," *Manitoba Free Press*, 1 July 1905, 27. Campbell's efforts to thwart the fund-raising are clear from comments in letters he wrote to W. M. Doull, 17 March, 13 April, and 15 September 1903, copies in Letterbook #7, held at West Kootenay Power's South Slocan archives.

[13] "Justice's Justice," *Nelson Tribune*, 8 August 1905, 1. For coverage of the trial, see "An Ominous Incident," *Nelson Daily News*, 13 July 1905, 3; "Power Company's Case," *Nelson Daily News*, 14 July 1905, 5; "Some Rock Will Float," *Nelson Daily News*, 15 July 1905, 5; "Practice and Theory," *Nelson Daily News*, 16 July 1905, 3; "Evidence Now All In," *Nelson Daily News*, 18 July 1905, 3.

[14] See A. A. den Otter, *Civilizing the West: The Galts and the Development of Western Canada*, Edmonton: University of Alberta Press, 1982, 156-57. For a general overview of transportation in the Kootenays, see John A. Eagle, *The Canadian Pacific Railway and the Development of Western Canada, 1896–1914*, Kingston, Montreal, London: McGill-Queen's University Press, 1989; Robert D. Turner, *West of the Great Divide: The Canadian Pacific Railway in British Columbia, 1880–1986*, Victoria: Sono Nis Press, 1987; and Robert D. Turner and David S. Wilkie, *The Skyline Limited: The Kaslo and Slocan Railway*, Victoria: Sono Nis Press, 1994.

15 See for example his comments in "Vancouver to Rossland," *Weekly News-Advertiser* (Vancouver), 21 October 1896, 8. (The article is an interview with Van Horne, concerning the construction of the Crowsnest Pass railway and its possible extension to the coast from the Kootenays.)

16 See "The Railway War at Sandon," *The Miner* (Nelson), 21 December 1895, 1; "Fun at Sandon," *Kaslo Claim*, 21 December 1895, 1. Robert Turner provides a very good account of this episode in Turner and Wilkie, *The Skyline Limited*, 61-65.

17 See for example the comments in British Columbia, *Annual Report of the Minister of Mines for 1905*, J25, quoted below. (These annual reports are printed in the *Sessional Papers* of the following year, so that the 1905 report appears in the *Sessional Papers* for 1906.)

18 British Columbia, *Annual Report of the Minister of Mines for 1905*, J175. This phrase was used again the following year: *Annual Report of the Minister of Mines for 1906*, H160.

19 British Columbia, *Annual Report of the Minister of Mines for 1909*, K131. For descriptions of Granby's mass mining techniques, see C. M. Campbell, "Granby Mining Methods," *Journal of the Canadian Mining Institute*, Vol. 11 (1908): 392-406 (reprinted in the *Engineering and Mining Journal*, Vol. 87 [30 January 1909]: 252-56); A. W. B. Hodges, "Handling Three Thousand Tons of Ore per Day at the Granby Mines and Smelter, Phoenix and Grand Forks, BC," *Journal of the Canadian Mining Institute*, Vol. 11 (1908): 407-14; and Roy Hutchins Allen, "Mines of the Granby Consolidated, Phoenix, BC," *Engineering and Mining Journal*, Vol. 88 (25 December 1909): 1260-62. For a useful historical summery of the Granby company's operations in the Boundary, see L. E. Carter, "Granby: Seventy-five Years of British Columbia Copper Mining," BA Hons Essay (Geography), Simon Fraser University, 1979, 17-24.

20 "CPR Is Outwitted," *Victoria Daily Colonist*, 15 July 1904, 1; cf. "Reported Control of Granby Mines," *Victoria Daily Colonist*, 14 July 1904, 6. It is difficult to ascertain if in fact Hill did control the Granby property at this point. John Fahey suggests that he did not (see *Shaping Spokane: Jay P. Graves and His Times*, Seattle: University of Washington Press, 1994, 20-21, 77-79), although his surviving correspondence files in the James J. Hill Library (St. Paul, Minnesota) indicate that he certainly exercised a good deal of influence on affairs of the company after 1904.

21 John Fahey points out that "the Great Northern between 1904 and 1910 [carried] three-fifths of the freight of the Boundary" (*Shaping Spokane*, 80). For the CPR's interest in the Boundary, see *Shaping Spokane*, 10, 15 81.

22 Aldridge to Sir Thomas Shaughnessy, 1 November 1904, Cominco Historical Files, Microfilm Reel #8, Add. Mss. 2500, BCARS.

23 For a description of these events, see Mouat, *Roaring Days*, 130-43.

24 James J. Hill to E. T. Nichols (Confidential), 7 June 1905, James J. Hill Letterbook, Hill Papers. A. J. Macmillan worked on Hill's behalf in Britain, and one can follow his activities through correspondence held in a file entitled "Merger of the Le Roi Mining Company et al., 1905–1906" (File 172, Box 12, in the papers of Crowsnest Resources Limited, M1561, Glenbow Archives).

25 Aldridge to C. A. Molson, 4 January 1906, Cominco Historical Files, Microfilm Reel #8, Add. Mss. 2500, BCARS.

26 *Ibid.* The St. Eugene mine was a silver lead property in Moyie, in East Kootenay.

[27] W. H. Aldridge to Sir Thomas Shaughnessy, telegram, 21 December 1904, held in RG2.A.A. #77128, Canadian Pacific Archives, Montreal.

[28] Although the CPR originally assumed that the amount that would be raised by the bond issue was a million dollars, it was subsequently raised to one and a half million dollars.

[29] Petition No. 20A, *Journals of the Legislative Assembly of the Province of British Columbia*, 1906, xli, submitted by Samuel S. Fowler, President of the Cascade Water, Power and Light Company, Limited, and dated 31 January 1906.

[30] A transcript of the proceedings, which ran from 7 to 13 February 1906, survives in a filing cabinet in West Kootenay Power's South Slocan Archives.

[31] Lorne Campbell to E. P. Davis, 23 May 1906, in West Kootenay Power & Light Company, Limited papers, file #7, Drawer #23, Cabinet #14, Rossland Historical Museum. For the charter of the South Kootenay Water-power Company, see the *Journals of the Legislative Assembly of the Province of British Columbia*, Vol. XXVI, 1897, xix.

[32] Details of the extraordinary general meeting of the South Kootenay Water Power Company, held in Rossland on 17 December 1906, are from a clipping held in historical file #11, South Slocan archives.

[33] "Cascade Water Power and Light Company, Limited," *The Mining Record*, Vol. 14, #11 (November 1907), 442.

[34] A copy of the brochure survives in the Cominco Papers, in Box 27, File 5, Add. Mss. 15 (held at the BC Archives in Victoria).

[35] Lorne Campbell to W. M. Doull, 16 April 1908, in Letterbook #14, held at West Kootenay Power's South Slocan archives.

CHAPTER FOUR

The Needs of War, the Need for Power: *The Trail Smelter, Zinc Production, and West Kootenay Power*

With the extension of service into the Boundary District and the completion of the new Bonnington plant by 1907, West Kootenay Power's sphere of operations included the most active mining regions of southeastern British Columbia. However, the expansion of both its capacity and service area did not change the main thrust of the company's business: its prosperity remained firmly tied to the resource industry whose mines and smelters consumed most of its electricity.

West Kootenay Power's most important customer was close to the site of its original Rossland operations. Cominco's smelter in Trail continued to grow, requiring more and more energy, and a close relationship began to develop between it and West Kootenay Power. In 1906, both were subsidiaries of the Canadian Pacific Railway. Ten years later Cominco took over the CPR's majority shareholding in West Kootenay Power, retaining ownership for the next seventy years. Naturally this had a considerable impact on West Kootenay Power: its expansion during the First World War and the twenty years that followed was largely in response to Cominco's needs. The electrical industry's growth in the region up until the Second World War reflected developments at Cominco.

Cominco only gradually emerged to become the dominant company in southeastern British Columbia. Its first years were scarcely promising ones. The company's Rossland mines were not rich producers and the other mine in the original Cominco group — the St. Eugene in Moyie — soon turned out to be on its last legs. The depletion of the St. Eugene led Cominco to lease, and later buy, the Sullivan mine in Kimberley. This proved a wonderful stroke of luck since the Sullivan ore was to become the major source of Cominco's wealth. Such an outcome was far from obvious at the time of the purchase, however. The Sullivan did not produce rich ore like the bonanza mines of the nineteenth century, nor could its ore be readily treated and smelted like the copper ore of the Boundary region. Indeed, the motive for the Cominco purchase in 1910 was simply to guarantee the Trail plant an adequate supply of lead ore, a necessary component in other smelting operations. Only sophisticated metallurgical processes could unlock the Sullivan's complex low grade lead-zinc ore and in 1910 no one in North America had mastered such techniques. Although it did not prove easy, once Cominco devised a satisfactory treatment method for the Sullivan ore, its future was bright indeed. So too was West Kootenay Power's, since the production of lead and zinc from the Sullivan ore required substantial amounts of electricity. The next three dams along the Kootenay River would be built in response to Trail's growing demand for power.

The story begins, however, with a crisis that overtook the Slocan mining industry in the early 1900s. It all had to do with the presence of zinc in lead ore. For many of Slocan's mine owners and managers, this was "the zinc problem." At the root of the problem lay the fact that the nature of the ore changed as mining operations went deeper underground. Zinc became a new and unwelcome component of the area's

The Zinc Problem

Unfortunately for the owner of mixed ores of lead and zinc, each metal strongly interferes in the process used for producing the other. When zinc ore goes into a blast furnace along with lead ore, the zinc is reduced as well as the lead, but the zinc gas at once burns in the blast to zinc oxide, which is an infusible solid and eventually chokes up the furnace. . . .

On the other hand lead in a zinc retort charge leads to equal difficulties. Part of the lead is carried over with the zinc and contaminates the spelter. Another part, as lead oxide, melts and running down through the charge to the bottom or hottest part of the retort, there combines with the clay of the retort to form slag, developing holes and destroying the retort.

– *Nelson Daily News*, 17 December 1908

silver-lead ore. And this was a problem because traditional smelting methods could not treat low-grade ores containing both lead and zinc.

"More and more the situation as regards lead mining in British Columbia," explained a government official, "becomes involved with the question of the disposal of the zinc contents of the ores."[1] Difficulties with zinc had closed many of the once-rich silver-lead mines of the Slocan by 1904 and a frantic search began for a method of treatment that would enable profitable operations to resume. If an adequate method of treatment could be implemented, zinc would quickly become a valuable asset instead of a crippling liability.

Europe was the acknowledged leader in zinc metallurgy; the metal had been mined and treated there for centuries.[2] As early as 20 BC, Romans were making brass out of copper and zinc ore and using it in their coinage, and commercial production of zinc began in the eighteenth century. Europeans roasted zinc ore in horizontal retorts and then recovered the pure metal by distillation. The drawback with this method — which was known as "the Belgian process" — was that it was time-consuming, labour-intensive, and effective only with relatively pure and high-grade zinc ores.[3] Low-grade lead-zinc ores such as those common in southeastern British Columbia could not be treated by this method. This became painfully clear when European investors backed an ambitious undertaking to erect a traditional zinc smelter in western Canada.

A European mining engineer toured the Kootenays in 1904. Constant Fernau concluded that the zinc problem was in fact a promising business opportunity and announced an ambitious and wide-ranging plan to treat the zinc ore of the province.[4] He would build a concentrator for the ore, a zinc smelter, and possibly a lead smelter as well. Confident that this venture would represent "another advance in the great progressive strides [the country] is now making," Fernau happily predicted that "The general result will be increased profits for the mine-owner, increased traffic for the railways, the employment of a larger number of men, an increase in the output, and an addition to the general prosperity of this section."[5] He wrote in a similar vein to the prime minister, although his real motive was to appeal for government assistance.[6]

Fernau did not secure government support but his Canadian Metal Company Limited managed to raise a million dollars in paid-up share capital from some five hundred French investors. These people were mostly doctors, lawyers, clerics, and widows, who typically invested the equivalent of a few hundred dollars.[7] The company soon constructed a zinc smelter in the Crowsnest Pass (at Frank, Alberta) and purchased the Bluebell mine, a silver-lead-zinc property on the east shore of Kootenay Lake.[8] The company certainly had a cosmopolitan approach: the plan was to smelt zinc ore from British Columbia in a plant financed by French capital, fuelled by Alberta coal, relying on Belgian technology, and employing American labour.[9] The Frank smelter — built during the spring of 1905, reputedly at a cost of $300,000 — produced its first zinc in early June of the following year.[10] Unfortunately for the French investors, the smelter was not the success that Fernau had predicted. A series of problems plagued the plant and it closed down after operating for only a brief time.[11]

The failure of the Frank smelter did not discourage others from trying to solve the zinc problem. A group of Nelson business people also attempted to find a way to treat the province's zinc ore. Electricity assumed a central part in their plans. Smelting operations needed a constant source of energy, the reason that Fernau had located his smelter in Frank, in the centre of a region rich in coal. The Belgian process needed two tons of coal to smelt a ton of zinc concentrate; thus, it was far cheaper to send the concentrates to the coal rather than the other way around. The Nelson group was confident that zinc could be smelted electrically, which would enable it to utilise the Kootenays' hydro-electric power. Like Fernau, members of the group appealed to various levels of government for financial and other assistance. In 1905, they persuaded the federal government to appoint a commission "to Investigate the Zinc Resources of British Columbia and the Conditions Affecting Their Exploitation." The commission — headed by the leading authority on zinc — took a year to produce an exhaustive study of

Words of Caution

. . . to anyone contemplating zinc smelting in the Canadian West I would emphasize again the necessity of being content with disappointing results for a considerable period while a sufficiently skilful working force is being secured. . . .

The process of zinc smelting is theoretically one of the simplest of all metallurgical processes. Practically it is one of the most difficult. This is because of the variety of steps that the ore must go through, the carefulness with which many details must be observed, and the necessity for carrying out the reduction of the ore in small retorts, made of imperfect and fragile material. No one should think of constructing a zinc smelter without employing a metallurgist skilled in the art.

– *Report of the Commission*, 1906

BCARS B-08525

This somewhat surreal photograph shows one of the blowers in Granby's Grand Forks smelter, driven by the electrical motor visible in the right foreground. Granby depended on West Kootenay Power for its electrical energy.

the topic, a four hundred page book with maps, photographs, statistics, and chemical analyses of the varieties of zinc ore. The commission held little hope for treating the province's zinc ore by any known method.[12]

The Nelson group did not wait to hear this gloomy prognosis from the commission that was largely its own creation. In 1905, the same year as the appointment of the zinc commission, the group launched the Canada Zinc Company, Limited, intending to smelt zinc using a patented electrical process. The Commission's report specifically rejected such a method as unsuitable to conditions in the province.[13]

Unlike the French backers of the Canadian Metal Company, the group behind the Canada Zinc Company included experienced mining engineers and business people, as well as the region's federal MP, W. A. Galliher. Most had lived in British Columbia and were familiar with the industry's problems.[14] The twenty shareholders each contributed a thousand dollars, the cost of a single share. These backers also possessed some political influence, persuading the provincial government in 1908 to advance a $10,000 loan.[15] The company built an extensive plant a mile outside of Nelson, complete with its patented electric smelting furnace, where it hoped to pioneer a commercial method of treating British Columbia's zinc ore.[16] Successful operations would require an ample supply of electricity.

West Kootenay Power was very interested in providing the new company with electricity. The City of Nelson's new Bonnington plant was having problems, although civic politicians were anxiously trying to resolve them. Lorne Campbell was not one to let an opportunity slip by; he assured an official with the Canada Zinc Company that "you do not need to feel at all anxious for we have agreed to take care of you, and . . . help you out in every way we possibly can."[17] Meanwhile, West Kootenay Power and the city continued jousting with each other. The difficulties with the City's new power plant meant that it had to purchase electricity from the company. The negotiations were not always smooth, making city representatives more determined than ever to have Nelson independent of the company.[18]

BCARS A-06635

The environment underground in the Granby mine was very different from the company's well-organized surface plants. The photograph shows an electrical locomotive, its skips being readied to take ore from an overhead chute.

> ### *The Power Plant*
>
> When we entered office [in January 1908] we inspected the [city's new Bonnington] plant and found it not only in a very uncompleted condition, but had it been so left there was every chance and probability of the entire investment becoming a wreck. . . .
>
> Had not, therefore, this piece of new construction been completed, you today would in all probability have been buying power from the West Kootenay Power & Light Co. at $45 per horse power, and would have been suffering the humiliation and great loss of a wrecked plant. . . .
>
> Nelson is and must be the "Electric City." No city in Canada is better lighted, and none has more liberal prices for electric current. Watch this industry with the greatest care and employ none but the most competent and thoroughly honest engineers. Guard against civic politicians who would try to trifle with your light and power rates.
>
> – *Nelson Daily News*, 2 January 1909

The Canada Zinc Company appeared to fulfil the promise of municipal ownership. Here was an example of industry coming to the city, taking advantage of the ready supply of electricity, and in the process contributing to the region's growth and prosperity. Unfortunately, it proved a little too good to be true. Despite considerable investment in terms of money, time, and expertise, the company's Nelson plant was not successful and it ceased operations at the end of 1908. Then the shareholders helped to organize another appeal to Ottawa. In November 1909, "the Zinc Producers of East and West Kootenay" requested further assistance from the federal government.[19] Their petition (submitted to the federal Minister of Mines by a Canada Zinc Company shareholder) described the zinc problem in British Columbia at some length, and claimed that the Canada Zinc Company had spent some $120,000 in its efforts to perfect electric smelting. This expenditure, they explained, had exhausted the financial resources of the company and only further government help—they estimated $20,000 to $25,000—could bring the costly experiments to a successful conclusion. But the federal government was reluctant to spend more money on what appeared to be a losing proposition. And it was receiving mixed messages. One of the prime minister's correspondents, for example, cautioned him that the real motive for the Nelson business people's interest in electric smelting was their desire to find a market for the hydro-electric power produced by the city's power plant.[20] The director of the Mines Branch in Ottawa suggested a thorough investigation of various new processes for treating zinc rather than simply putting additional money into the experimental electrical plant in Nelson. Parliament agreed and voted to set aside $50,000 for zinc research.[21]

The industry certainly appeared to need help. A former president of the Canadian Manufacturers' Association described the depressed state of the mining industry in the Kootenays when he passed through Nelson in 1910. In a newspaper interview he repeated a familiar refrain: "The solution of the zinc smelting problem . . . [is] the matter of chief importance to the mining industry of British Columbia." This attitude may have been no more than self-interest, a response to the conditions he found at the Slocan mine in which he had invested. But his comments suggest a genuine concern: "none can pass through the once thriving cities of New Denver, Slocan City, and Silverton without seeing the unfortunate results of the depression caused by the zinc reduction problem."[22] Help, however, was already on its way.

Andrew Gordon French arrived in Victoria in 1909. French hailed from Glasgow and had worked as a metallurgist in Swansea, Wales.[23] People differed widely in their descriptions of him. A senior official with Cominco, engaged in a lawsuit with French's company, claimed he was "a rather clever and unscrupulous man . . . [who] apparently

KOOTENAY MUSEUM ASSOCIATION AND HISTORICAL SOCIETY, 704

The much-delayed opening of the City of Nelson's power plant at Upper Bonnington was the occasion for this dignified photo opportunity in 1910.

cribbed all his knowledge of zinc from the earlier patents and processes."[24] On the other hand, one of French's financial partners assured Prime Minister Laurier in 1911 that he was "a gentleman of great scientific ability and well known among Mining Men throughout the world."[25] Several years later the premier of British Columbia, responding to a direct question from the prime minister as to French's character, summed him up in one word: "eccentric."[26]

Just how much French knew about zinc metallurgy became highly controversial. He knew enough, certainly, to appreciate the importance of pioneering a successful means of extracting zinc from low grade ores, as well as the potential financial reward for patenting such a method. Not long after arriving in the province, French announced that he had invented a zinc process and persuaded Victoria business people to organise a company to provide him with financial backing, French's Complex Ore Reduction Company.

French announced in September 1910 that he would re-locate to Nelson, to perfect his process in a spot close to the province's troublesome zinc ores and close as well to a convenient source of power. The city of Nelson welcomed him, giving French a lease on its old Cottonwood powerhouse for his experimental plant.

In Operation By Christmas

"I cannot give any definite date as to the starting up of the small zinc smelter which is to demonstrate the success of my method of reducing low grade zinc ores, but unless something unforeseen occurs we will be in full running order by Christmas," said A. Gordon French, the prominent old country metallurgical chemist. . . .

"I am now engaged upon the question of the necessary machinery for the smelter which will be installed, as has been already stated, in the old city power house. . . ."

— *Nelson Daily News*, 6 October 1910

A year later French successfully produced metallic zinc, using a process known variously as the Létrange process, the wet extraction process, or the electrolytic process. Although very different from the electric smelting earlier attempted by the Canada Zinc Company—and technically the French process was refining, not smelting—it too required a substantial amount of electricity. A three stage procedure, the process involved roasting zinc ore, then leaching it in solution, and finally recovering the metallic zinc by electrolysis. Metallurgists were familiar with the general approach but regarded it as only theoretically feasible. It had been discredited in the 1890s when the efforts of a British company to establish it on a commercial basis ended in costly and well-publicized failure.[27]

French's apparent success attracted much attention. The Provincial Mineralogist paid a visit to his Nelson plant in the autumn of 1911 and later published a thorough report.[28] Cominco was also very interested in French's work. In November 1911 Cominco's general manager, R. H. Stewart, met French and discussed the possibility of using French's process to treat Sullivan ore. Stewart returned to Nelson a few months later with two other members of Cominco's staff. One of them stayed in Nelson for three months, studying the process at first hand. He must have filed a favourable report for in late May 1912 Cominco and French signed a contract giving the company exclusive rights to French's process. Once the process was successfully in operation, Cominco would pay a lump sum of $200,000 as well as a royalty on zinc production.[29]

WEST KOOTENAY POWER – TRAIL OFFICE

Lorne Campbell poses in his office, March 1911. With the swivel chair, roll-top desk, telephone, and secretary ready to take dictation, he seems the very model of a modern business manager.

In addition, Cominco agreed to build a zinc plant at Trail within a year with a capacity of one ton a day. The company changed its plans several times, however, and the plant was not ready until May 1914. French's son, Thomas, then moved to Trail to get the zinc plant up and running. Cominco, however, became increasingly dissatisfied with Thomas French's work and in January 1915 it cancelled the contract with French's Complex Ore Reduction Company.

By this time, of course, the First World War had begun. With war came a new demand for base metals. The government of Canada had agreed to supply munitions to the British as part of the Canadian war effort. The fact that the country had no refineries capable of producing either copper or zinc — the constituent metals of brass — meant that Canada's munitions factories had to import refined copper and zinc from the United States. For the first three years of the war, however, the United States was neutral. Reliance on a non-combatant country for essential war materials caused concern in Ottawa.

An Anxious Time

Zinc Shortage. – The great shortage of zinc at that time was serious. . . . Canada had produced for some years the concentrates of zinc and copper matte, but had no facilities to refine them, and sent them to the USA for that purpose.

German Control. – The Germans controlled the zinc and copper markets of the USA. It was known that some American zinc smelters refused to supply zinc to Britain. The USA were Canada's only source of supply. Britain could not help her. She had to import these metals for her own requirements. . . . The large increase in the demand for cartridge case material made it imperative that the source of supply should be assured. The Shell Committee entertained fears that public opinion and German interests in the USA might force the Government to place an embargo on these metals.

– *The History of Munitions Supply in Canada 1914–1918*

The federal government appointed a commission "to inquire as to the possibility of refining copper and producing metallic zinc in Canada."[30] This second zinc commission — which, like the first, produced a substantial report of many pages — concluded that Cominco's smelter at Trail was probably best situated to produce zinc and copper. The smelter was Canadian-owned, it was close to ore supplies and the necessary hydro-electric power, unencumbered by any binding contracts, and its staff included skilled metallurgists.[31] In June 1915, the zinc commission recommended that Cominco "be asked to enlarge their [experimental] plant to produce 25 tons of metallic zinc per day. . . . The amount of money for such extension to be advanced either by the Shell Committee or the Canadian Government. . . ."[32] Ottawa was reluctant to make such an outright capital advance but apparently decided that war-time conditions justified the move.[33]

After backing out of its arrangement with French, Cominco continued to experiment with zinc extraction processes, with the added incentives of sky-rocketing prices and high demand brought on by the onset of the war. The company's metallurgists concluded that the basic stages that French had employed in his process — roasting, leaching and electrolysis — did offer the best means of treating the Sullivan ore. But perfecting this process and then engaging in large-scale commercial production was no easy matter. If the company received government money to construct a zinc refinery, in exchange for a guarantee to produce so many tons of zinc per day, it was making a commitment that some Cominco staff felt that it might not be able to honour. They did not feel that their grasp of the process was such that they ought to make such a pledge.

Cominco's chances of success

Referring to the proposition of building a zinc plant to make twenty to twenty-five tons of metallic zinc per day, — I have been going into this matter pretty fully since you left here and it seems to me there is only about one chance in a hundred of our being able to make twenty tons of zinc a day at the end of six months. . . .

Excepting for the possible urgent need for zinc, a lot more work should be done before commencing operations on a large scale. . . . I don't think we ought to promise to produce any stated tonnage in any stated time.

— **S. G. BLAYLOCK**, 20 July 1915

Cominco's directors, however, were determined that the company should do all it could to assist in the war effort. This, they felt, was a patriotic duty. They overrode the objections and ordered a twenty-five ton zinc plant for Trail.[34]

A series of letters and telegrams from the autumn of 1915 demonstrates the intimate relationship between the CPR, Cominco, and West Kootenay Power, as well as the fundamental importance of abundant cheap electricity for successful zinc treatment. Cominco had little trouble in persuading Campbell to expand the generating capability of the Number Two plant at Upper Bonnington. Campbell also oversaw the construction of two new pole lines to the Trail smelter, each carrying 60,000 volt circuits. J. J. Warren, Cominco's managing director, kept CPR president Shaughnessy apprised of developments. In turn, Shaughnessy relayed requests and information on to Hosmer, president of West Kootenay Power. Hosmer would then contact Campbell. In all the to-ing and fro-ing, however, the respective companies never lost sight of their own interests. For example, when Hosmer advised Campbell that Cominco would need a lot more power if the zinc process proved successful, he added, "do not commit yourself as regards the price that you will charge for such additional power as they may want, as we think we should get an increased price, or, in other words, that we should participate in their extraordinary profits."[35]

The need for power

The Consolidated Mining & Smelting Company are about to close a contract for a considerable quantity of zinc, which will, naturally, make an increased demand for power on the West Kootenay Company.

I think that Mr. Campbell feels that, as a safeguard, another generator should be installed without delay. Will you kindly communicate with him on the subject and arrange.

If necessary, means will be found for assisting in providing the requisite funds for the purpose. . . .

— **T. G. SHAUGHNESSY** to **C. R. HOSMER**, 23 December 1915

Construction of Cominco's new zinc plant began in the autumn of 1915. The company's commitment to electrolytic zinc production led inexorably to a far more intimate relationship between it and West Kootenay Power; and, in January 1916, Cominco acquired the company from the CPR. At first, however, it seemed as if the misgivings of Cominco's staff were amply justified, and serious problems continued to dog their tentative method of zinc treatment. Only the metal's high price allowed zinc production to remain relatively profitable, but this reflected war-time conditions and would not last forever. The company's metallurgists knew that for Cominco to achieve long-term success as a zinc producer, ore from its Sullivan mine needed to be in a purer state when it arrived at the Trail complex.[36] As luck would have

ROSSLAND HISTORICAL MUSEUM 597-21

Relaxing along the Kootenay River, in the calm before the First World War. West Kootenay Power's superintendent at Upper Bonnington appears more at ease than Lorne Campbell: well-protected from any solar radiation, he and his family pose for the camera before going for a drive in the open air.

it, a method that would allow this to happen had already been developed in Australia.

The development of the flotation process, a crucial advance in ore separation, had come a dozen years earlier in Broken Hill, New South Wales, site of one of the world's largest silver-lead-zinc deposits.[37] Flotation offered a solution to the problem of treating the Sullivan ore, since if it could be applied successfully, it would concentrate the crude ore—turning it into a relatively high-grade product—before shipment to Trail for further refining. This was a considerable advantage since Cominco's electrolytic process was far more efficient with higher grades of zinc concentrates; the problems with the process were largely a consequence of the low grade of the Sullivan ore. Sensing a solution to their problems, the research staff at Cominco quickly turned their attention to flotation.

In 1917, Selwyn Blaylock, Cominco's manager at Trail, persuaded R. W. Diamond, a Canadian then working for the Anaconda Copper

Mining Company, to return to Canada and help Cominco find a way to mill the Sullivan ore. (Many years later, Diamond would succeed Lorne Campbell as president of West Kootenay Power.[38]) Diamond's experience in Butte with Anaconda had given him a good deal of experience with flotation, and after a period of trial and error he developed a customized process capable of treating the Sullivan's low-grade ore.[39] This took time, however, and the Sullivan concentrator at Kimberley did not go into operation until the summer of 1923. The event had a dramatic impact on Cominco's profitability: having mastered a means to treat the ore as well as a way to refine it, the company could reap the full benefits of the massive reserves of Sullivan ore.

Success attracts attention and not always the sort of attention that one wants. Not long after the new Sullivan concentrator went into production, Andrew Gordon French's company took Cominco to court

KOOTENAY MUSEUM ASSOCIATION AND HISTORICAL SOCIETY, 1458

Further down the river, the log cabin at Slocan Pool was a well-known focal point, a good spot to fish and picnic. Here a group poses formally with the catch of the day; hopefully they brought packed lunches as well.

for patent infringement. French's Complex Ore Reduction Company of Canada claimed that Cominco was using the patented French process in the Trail smelter, that as a result Cominco had become a very prosperous business, and that consequently Cominco should pay French's company some eight million dollars compensation.[40] It initiated legal proceedings in July 1924, although the issue had been raised as early as 1917.[41]

Cominco regarded the court action "as more or less of a blackmail proposition and that they [the owners of the French patents] are out for a settlement out of court."[42] Firmly rejecting such overtures, Cominco demanded that the issue be settled in the courts. "I feel very strongly that the case should be fought out," wrote the company's feisty general manager, ". . . the reputation of the Company and the Officials of the Company is worth far more than the cost of defending the action would be."[43] Not long after French's Complex Ore Reduction Company began proceedings against it, Cominco created a new company, the Electrolytic Zinc Process Company, a partnership with Anaconda, the American mining giant. Like Cominco, Anaconda had developed an electrolytic process for treating zinc during the First World War and it too could be involved in litigation should French's action prove successful. In mid December 1924, Electrolytic Zinc launched a counter suit against French's company in the Exchequer Court in Montreal, aiming to have the key French patent declared invalid.[44]

This began a lengthy courtroom saga. Two years passed before the actual trial began, which lasted from early February to early March, 1927.[45] "The trial has been long!" exclaimed the Hon. Mr. Justice Audette when he pronounced judgment; "The trial has been long, but it has been exhaustive and very ably argued."[46] He had heard complex and often contradictory testimony on the evolution of zinc metallurgy from the 1870s to the 1920s: "the history of the ore from beginning to end," he noted without enthusiasm at one stage.[47] This history of past practice, however, was the central issue since the case hinged on whether or not French's patent described an original process.

Cominco and Anaconda had a strong case to argue and its effectiveness was considerably enhanced when the two companies secured the service of zinc expert Walter Ingalls. He had headed the 1906 Zinc Commission in British Columbia and also worked for the federal government on the "zinc problem." The judge found Ingalls "a witness of unusual knowledge and experience" and — perhaps swayed by his testimony — declared French's patent invalid.[48] He also ordered the defendant, the French's Complex Ore Reduction Company of Canada, to pay all the legal costs. Although the company subsequently appealed the Exchequer Court decision to the Supreme Court of Canada, the appeal was dismissed.[49]

The judgment in the case did more than ensure that Cominco and Anaconda would not be forced to make royalty payments or face similar restrictions. The decision also strengthened Cominco's position as BC's leading smelter. (The Granby smelter in Grand Forks had closed down in 1919 and the company's operations moved to Anyox on the province's north coast.) As with West Kootenay Power, the fact that the CPR was the majority shareholder in Cominco attracted charges of monopoly and corporate domination. Just as Houston and others in Nelson railed against West Kootenay Power as a force that threatened local initiative, a number of British Columbia's smaller mining companies came to resent Cominco's dominant position in the province's mining industry. They linked the need for government help in finding a solution to the zinc problem with their ability to compete with large corporations, such as Cominco. Thus when the federal government decided to assist Cominco establish its zinc plant in 1915, the Mines Committee of the Vancouver Board of Trade protested the move in an indignant letter to the BC premier, Richard McBride.[50]

A word to the Premier

The Trail Smelter Zinc works being erected is a private concern and will not smelt zinc ore from outside mines without a very large profit. This is exactly what is crushing the life out of the small mining companies that don't own a smelter. We consider it to be the duty of your Government to make arrangement [so] that the individual mines and companies, who have not the capital to erect a Smelting Plant of their own, [will be able to] get their ore treated at as low a rate as any of the Mining and Smelting Companies you have named treat their own ores.

— **THE VANCOUVER BOARD OF TRADE** to Premier McBride, 13 November 1915

In the spring of 1917, the province's new Liberal government enacted legislation "for the Establishment of Public Sampling and Concentrating Plants, Custom Smelters and Refineries," a move that was likely a response to such pressure. Similar moves were apparent in Ottawa too: when the House of Commons discussed giving aid to Cominco in mid-1918, one MP angrily charged the federal government with being "very anxious to create a monopoly." Cominco, he claimed,

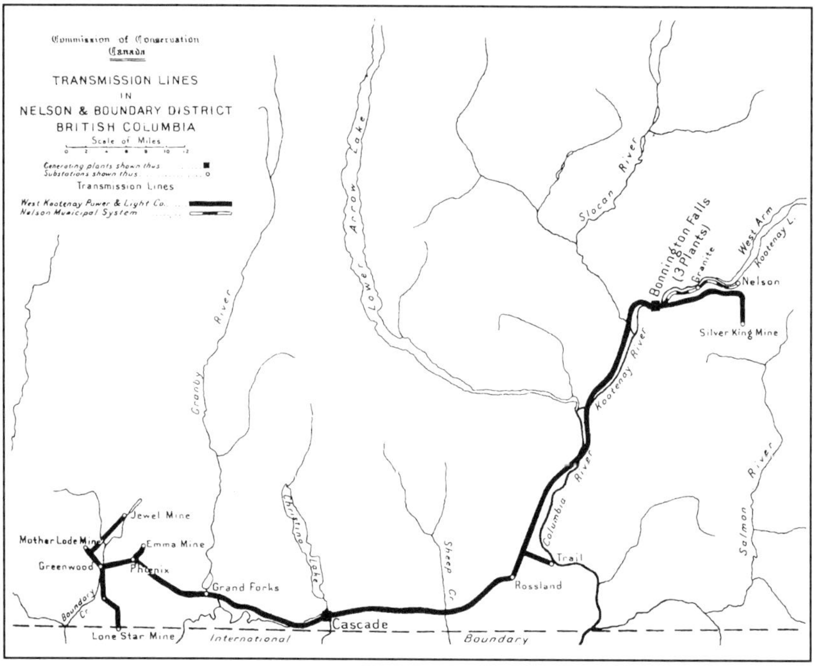

The extent of West Kootenay Power's transmission lines at the end of the First World War. This map appeared in a federal government publication (*Electric Generation and Distribution in Canada*), published in 1918.

"formed a great monopoly in the province of British Columbia by reason of having this smelter, which is going to be supported by the Government. . . . The Consolidated Mining Company have [*sic*] put all the small miners out of business, and are now in possession, practically speaking, of the whole mining system of the Kootenay district."[51] Pressure for some form of government intervention continued. In the autumn of 1919, for example, people in the Kootenays lobbied long and hard for an experimental ore treatment plant for the district.

Letters and appeals for an experimental plant continued for three years—all of them repeating a familiar litany and indicative of the growing breach between Cominco and the smaller operators—but nothing ever came of the initiative.[52]

Whatever the justice of the accusations of monopoly, Cominco did become the giant enterprise of the western mining industry. The company's success and longevity were not guaranteed, however, until the zinc problem had been solved. Once Cominco mastered electrolytic

And a telegram to the federal Minister of Mines

[Cominco] have for years been experimenting with the ore from their Sullivan Mine, expending several hundred thousand dollars towards that end. . . . It is self evident that whilst wealthy corporations . . . can afford to undertake the risk of making large expenditures in that direction, it is quite beyond the means of smaller concerns to do so. . . . Therefore the meeting [of the Nelson Board of Trade] was unanimous in claiming that it is the duty of the government to provide a suitable plant which will give the smaller concerns the necessary facilities for ascertaining the proper and most economical treatment for their complex ore. . . . Such a plant . . . would induce capital to the country and would do away with very much of the friction existing today between the shippers and the smelters.

— **FROM THE NELSON BOARD OF TRADE**, 7 October 1919

zinc refining and adopted the flotation process for its Sullivan ore, the share values of the company rose dramatically.[53] By 1941 the company was the most profitable ever to have operated in British Columbia and its dividends very nearly equalled those paid out by all other mining companies that had ever been active in the province. As late as 1966 a government publication could credit the company with having produced "nearly one half the metal wealth of British Columbia."[54]

Cominco's prosperity was based not only on its growing metallurgical prowess and its possession of the massive lead-zinc Sullivan ore deposit, but also on its access to abundant hydro-electricity. The company's interest in, and success with, zinc production served to underline West Kootenay Power's critical role in its operations. Cominco's growth and expansion would be possible only with a matching rise in West Kootenay Power's generating capacity.

NOTES

1 "Annual Report of Mr. G. O. Buchanan, Inspector of Lead Bounty Claims," Kaslo, BC, January 16, 1914, in CANADA *Sessional Papers*, Vol. 48, No. 7, 1914, 11. A petition by "the Zinc Producers of East and West Kootenay" provides an excellent account of the difficulties that mining companies faced as they encountered increasing amounts of zinc. See Louis Pratt to William Templeman, Minister of Mines, 18 November 1909, Laurier Papers, MG26-G, Vol. 598, #162335-342, National Archives of Canada [NAC hereafter].

2 For the history of zinc treatment, see J. M. Dawkins, *Zinc and Spelter: Notes on the Early History of Zinc . . .*, Oxford: Zinc Development Association, 1950; E. J. Cocks and

B. Walters, *A History of the Zinc Smelting Industry in Britain*, London: George G. Harrap & Co. Ltd., 1968; and P. T. Craddock, ed., *2000 Years of Zinc and Brass*, London: British Museum, 1990.

3 For a description of zinc smelting by the Belgian process, see W. H. Dennis, *A Hundred Years of Metallurgy*, Chicago: Aldine Publishing Co., 1964, 158-66.

4 "Zinc Plants for the Slocan," *Daily Province* (Vancouver), 25 February 1904, 4.

5 *Ibid.* See also "Reduction Plant," *Daily Province*, 8 March 1904, 2; "A Zinc Smelter," *Daily Province*, 31 August 1904, 2; "Zinc-Reducing Plant," *Daily Province*, 14 September 1905, 4.

6 Fernau & Co. to Laurier, (no date, but presumably early 1904, given its position in Laurier's correspondence files), MG26-G, Vol. 297, #80465h-80465t, NAC.

7 For a list of shareholders as of July 1906, see British Columbia, Dept. of Attorney General, Registrar of Companies, Company Registration Files, 1897 series, Microfilm Roll 7, B4422, GR1438, Canadian Metal Company Limited, File #1195, 51-61, British Columbia Archives and Records Service [BCARS hereafter].

8 For the smelter plans, see *Report of the Commission Appointed to Investigate the Zinc Resources of British Columbia and the Conditions Affecting their Exploitation*, Ottawa: Mines Branch, Dept. of the Interior, 1906, 63-64. For brief discussions of the Bluebell mine, see W. Donald Fraser, "Historical Geography of the Riondel-Gray Creek Region 1882–1940," BA Essay (Geography), UBC, 1964, esp. 25-27, and Craig Weir, "Bluebell The Storybook Mine," *Western Miner & Oil Review*, Vol. 33, No. 5 (May 1960), 26-31.

9 For a reference to the American labour used at the plant, see *A Record of the Investigation, Report and Subsequent Action of the Commission . . . To investigate the feasibility of Refining Copper and Producing Metallic Zinc on a Commercial Scale in the Dominion of Canada*, Ottawa: Imperial Munitions Board, 1916, 206.

10 "Refining Zinc—First Spelter Made in Canada Turned Out by the Frank Smelter," *Victoria Daily Times*, 13 June 1906, 3. The cost of $300,000 is given in "Investigation of Processes for the Reduction of Refractory Zinc Ores," *Summary Report of the Mines Branch of the Department of Mines . . . 1910*, Ottawa, 1910, 12.

11 One of the difficulties with the smelter may have been that its management was less than honest. In early 1906, manager Thomas Jones had been found guilty, along with A. C. Garde, of cheating a Vancouver man on the value of some shipments; the judge described it as a "businesslike arrangement between two men to cheat a third." ("Court Roasted Zinc Buyers," *Daily Province*, 22 February 1906, 5). In July 1906, the management of the company was changed considerably, presumably in response to problems with the smelter ("Western News," *Weekly Herald* [Calgary], 2 August 1906, 9). For a good analysis of the smelter's failure, see *A Record of the Investigation, Report and Subsequent Action of the Commission . . .*, Ottawa, 1916, 187, 206-07.

12 The activities and findings of the Commission are discussed at length in the official *Report of the Commission Appointed to Investigate the Zinc Resources of British Columbia and the Conditions Affecting Their Exploitation*, Ottawa: Mines Branch, Department of the Interior, 1906. For an interesting account of the genesis of the Commission, see "Zinc Mining in British Columbia," *Canadian Mining Journal*, Vol. 31, #9 (1 May 1910), 274. W. R. Ingalls, the person recruited to head the investigation, was then editor of the prestigious New York-based *Engineering and Mining Journal* and the annual publication, *The Mineral Industry*. He was also the author of two books on zinc metallurgy (*Production and Properties of Zinc . . .*, New York: Engineering and Mining

Journal, 1902, and *The Metallurgy of Zinc and Cadmium*, New York: McGraw-Hill Book Company, 1903). For evidence of Ingalls' lack of enthusiasm for zinc smelting in BC, see *Report . . .* , 55, 72, 133.

13 See, for example, *Report* . . . , 133. Electric smelting refers to a specific type of furnace; in essence it was the traditional Belgian process using a patented furnace fired by electricity. It was very different from the electrolytic method of zinc treatment subsequently developed by Cominco.

14 See the British Columbia, Dept. of Attorney General, Registrar of Companies, Company Registration Files, 1897 series, Microfilm Roll 7, B4422, GR1438, BCARS, Canada Zinc Company, File #1249, 23. This is the list of shareholders as of 26 November 1907, and gives the names of nineteen men and one woman as shareholders.

15 British Columbia, *Annual Report of the Minister of Mines for 1907*, 23.

16 The design of this furnace incorporated a patented process acquired from Frederick Snyder, an American engineer and inventor, whose patents were the key asset of the company. The process is described by Frederick T. Snyder in "Electric Zinc Smelting," *The Canadian Mining Journal*, Vol. 1, #16, New Series, 1 November 1907, 492-93. For the patents, see No. 111,050, "Ore Treatment," *The Canadian Patent Office Record*, Vol. 36-1, March, 1908, 731-32, and No. 111,673, "Process of and Apparatus for Treating Ores," *The Canadian Patent Office Record*, Vol. 36, No. 2, May, 1908, 1087-88. For a description of the Nelson operation, see "Plant of Canada Zinc Co. at Nelson," *Annual Report of the Minister of Mines for 1908*, 173-75.

17 Lorne Campbell to E. F. Miltenberger, 6 March 1908, in Letterbook #14, South Slocan Archives. These discussions had been going on for some time: sixteen months earlier, Campbell had outlined the terms on which West Kootenay Power would supply electricity to the company. See Lorne Campbell to Canada Zinc Company, 16 November 1906, in "West Kootenay Power and Light—General," File #1, Drawer #23, Cabinet #14, Rossland Historical Museum.

18 See for example "Power Plant, Lorne Campbell Makes an Offer to the City," *Nelson Daily News*, 4 February 1908, 1, 8; "Makes Terms, City's Arrangement With West Kootenay Co.," *Nelson Daily News*, 11 February 1908, 1; and "Civic Affairs," *Nelson Daily News*, 2 January 1909. The city's plant was not fully operational until the autumn of 1910; see "Power Asset, City Has Splendid Plant at Bonnington" *Nelson Daily News*, undated clipping, October 1910, in scrapbook, "West Kootenay Power and Light—General," File #1, Drawer #23, Cabinet #14, Rossland Historical Museum.

19 See "Investigation of Processes for the Reduction of Refractory Zinc Ores," *Summary Report of the Mines Branch of the Department of Mines . . . 1910*, Ottawa, 1910, 11-15, where the petition of the Zinc Producers of East and West Kootenay to Templeman, Minister of Mines, is reprinted. A copy of the original petition, dated 18 November 1909, survives in the Laurier Papers (MG26-G, Vol. 598, #162335-342, NAC).

20 See E. O. Windsor to Laurier, 24 December 1909, in Laurier Papers, MG26-G, Vol. 604, #164136, NAC.

21 See *Summary Report . . . 1910*, 13-15, and CANADA *Debates*, Session 1909–10, Vol. III, (21 March 1910), 5794-5801; also Vol. IV, (30 March 1910), 5935-41. The subsequent government investigation of appropriate methods for treating the zinc ores of the province, again headed by Walter Ingalls, went on for four years. In the spring of 1914 he announced that a private company seemed close to perfecting an electric process, and that in his view to continue the government research would simply be an un-

productive duplication of effort. See the report of Ingalls' speech in "Treatment of British Columbia Zinc Ores," *Canadian Mining Journal*, Vol. 35, No. 10 (15 May 1914), 334-36.

22 "Urges Solution of Zinc Smelter Problem," *Daily Province*, 10 October 1910, 18.

23 For brief discussions of French's background, see "Expert Will Study Ores in Kootenay," *Daily Province*, 2 September 1910, 8; "New Method for Smelting Zinc," *Victoria Daily Times*, 22 September 1910, 14; "The Zinc Problem," *The British Columbia Mining Record*, Vol. 17, No. 3 (October, 1911), 79-81; T. A. Rickard, "Electrolytic Zinc at Trail," *Mining and Scientific Press*, Vol. 113, 30 December 1916, 933; and the flattering self-portrait in A. G. French to J. O. Patenaude, 8 August 1911, in Laurier Papers, MG26-G, Vol. 689, #188710-714, NAC.

24 S. G. Blaylock to H. W. Gepp, 3 July 1924, Trail, Box 195, Add. Mss. 2500, BCARS.

25 J. O. Patenaude to Laurier, 8 August 1911, MG26-G, Vol. 689, #188707, NAC.

26 Richard McBride to Robert Borden, 23 January 1915, MG26-H, Vol. 198, #110429 (microfilm reel C-4392), NAC.

27 This was the Sulphide Corporation's attempt to treat the refractory sulphide ores of Broken Hill at Cockle Creek, New South Wales, in 1897–1898. For a description of this, see Edgar A. Ashcroft, "The Treatment of Broken Hill Sulphide Ores by Wet Extraction Processes, and the Electrolytic Deposition of Zinc," *Transactions of the Institution of Mining and Metallurgy*, Vol. VI (1898), 282-337; also Walter Renton Ingalls, "Progress in the Metallurgy of Zinc," *The Mineral Industry, Its Statistics, Technology and Trade*, Vol. VI (1898), 667-70; Ingalls, "The Treatment of Mixed Sulphide Ores," *The Mineral Industry, Its Statistics, Technology and Trade*, Vol. VII (1899), 743-44; and Ingalls, "The Ashcroft Process for the Treatment of Mixed Lead and Zinc Sulphide Ores," *Engineering and Mining Journal*, Vol. 66 (22 October 1898), 488-90.

28 Wm. Fleet Robertson, "French's Process For Separation of Zinc and Lead," in British Columbia, *Annual Report of the Minister of Mines for 1911*, 162-65. The records of the Provincial Mineralogist have "Papers relating to zinc, and Andrew Gordon French," in GR1097, File 4, BCARS; see esp. "French's Process," a five-page typescript with pencilled notes by A. G. French. Note as well the brief descriptions in *Annual Report of the Minister of Mines for 1910*, 103 and *The British Columbia Mining Record*, Vol. 17, No. 3 (October 1911), 79-81.

29 Plaintiff's Declaration, *French's Complex Ore Reduction Company of Canada, Ltd., Plaintiff vs Consolidated Mining & Smelting Co. of Canada, Ltd., Defendant*. No. 1804, Superior Court, Montreal, Quebec. June, 1925. Typescript copy held in File 1, Box 193, Add. Mss. 2500, BCARS.

30 David Carnegie, *The History of Munitions Supply in Canada 1914–1918*, London: Longmans, Green & Co., 1925, 60.

31 See *A Record of the Investigation, Report and Subsequent Action of the Commission . . . To investigate the feasibility of Refining Copper and Producing Metallic Zinc on a Commercial Scale in the Dominion of Canada*, Ottawa: Imperial Munitions Board, 1916, 6-8, 273, 278-80.

32 *Op. cit.*, 273.

33 For the government's misgivings, see the letter of Thomas White, Minister of Finance, to Carnegie, 7 July 1915, *op. cit.*, 275-76. Blaylock described the agreement reached with the government in his speech at the Revelstoke mining conference, 11 July 1918,

esp. 4-6 (a typescript copy is held in French's Complex Ore Reduction Company papers, Add. Mss. 2500, BCARS); see also the remarks of J. J. Warren, Managing Director, in the Cominco *Annual Report* of 1916.

34 See the letter from Cominco's managing director, James J. Warren, to Sam Hughes, Minister of Militia, 15 June 1915, reprinted in *1916 Zinc Commission*, 267-68.

35 C. R. Hosmer to L. A. Campbell, 15 December 1915, marked "Personal," file 20, South Slocan Archives.

36 For elaboration of this point, see R. W. Diamond, "The Development of the Treatment by Flotation of the Ore of the Sullivan Mine, Kimberley, BC 1917–1923," *Quarterly of the Colorado School of Mines*, Vol. 56 (July 1961), 521.

37 For an account of this discovery, see Jeremy Mouat, "The Development of the Flotation Process: Technological Change and the Genesis of Modern Mining, 1898–1914," *Australian Economic History Review*, Vol. 36, No. 1 (March 1996), 3-31. The report of the 1906 Zinc Commission devoted a few pages to flotation, noting apologetically that pressures of time and money prevented a really thorough investigation: see the *Report of the Commission Appointed to Investigate the Zinc Resources of British Columbia . . .*, 121-28, esp. 125. Another five years passed before the first North American flotation plant was built in Butte, Montana.

38 See Diamond's comments in Interview #7, 15 July 1975, 1-2, as well as the booklet that he wrote, *A Detailed Account of the Development of the Treatment by Flotation of the Ore of the Sullivan Mine, Kimberley, BC, 1917–1923*, Trail, 1961. Although based on his article in the *Quarterly of the Colorado School of Mines*, the booklet provides some additional details.

39 Diamond's article cited above provides a very good account of his research, but see also the jointly-authored piece, "The Development of the Sullivan Mine and Processes for the Treatment of its Ores," *Transactions of the Canadian Institute of Mining and Metallurgy*, Vol. XXVII (1924), 306-69.

40 Plaintiff's Declaration, *French's Complex Ore Reduction Company of Canada, Ltd., Plaintiff vs Consolidated Mining & Smelting Co. of Canada, Ltd., Defendant.* No. 1804, Superior Court, Montreal, Quebec, June 1925; typescript dated 25 July 1924; copy held in File 1, Box 193, Add. Mss. 2500, BCARS. Asserting that French had invented the electrolytic process used in the Trail smelter to treat Sullivan ore, the French's Complex Ore Reduction Company of Canada explained that "the present action, . . . [was] based on the infringement of . . . Patent [140402] by the Defendant" (Plaintiff's Answer to Plea, *French's Complex Ore Reduction Company of Canada, Ltd., Plaintiff vs Consolidated Mining & Smelting Co. of Canada, Ltd., Defendant.* No. 1804, Superior Court, Montreal, Quebec, June 1925; typescript dated 15 June, 1925; copy held in File 1, Box 193, Add. Mss. 2500, BCARS). Andrew Gordon French had transferred his patents to French's Complex Ore Reduction Company in October 1912. This company evidently reorganised, for in August 1923 French's patents were transferred to French's Complex Ore Reduction Company of Canada, Ltd. The latter company was based in Montreal.

41 In early 1917, for example, S. G. Blaylock (assistant General Manager at Trail) wrote to Cominco's Managing Director, responding to "the statements that the operations of the Trail zinc plant have infringed on French's patents." (Blaylock to J. J. Warren, 14 February 1917, Trail; Box 195, Add. Mss. 2500, BCARS.)

42 S. G. Blaylock to H. W. Gepp, 3 July 1924, Trail. Box 195, Add. Mss. 2500, BCARS.

43 S. G. Blaylock to J. J. Warren, 11 September 1924, Trail, Box 195, Add. Mss. 2500, BCARS.

44 Statement of Claim by Plaintiff, Brief No. 1, *Electrolytic Zinc Process Company, Plaintiff vs French's Complex Ore Reduction Company of Canada, Ltd., Defendant.* Filed in Exchequer Court of Canada, 19 December 1924; typescript copy held in File 7, Box 193, Add. Mss. 2500, BCARS.

45 This trial was the counter suit launched by Cominco and Anaconda (*Electrolytic Zinc Process Company vs French's Complex Ore Reduction Company of Canada, Ltd.*) concerning the validity of French's patent. The original suit involving infringement, *French's Complex Ore Reduction Company of Canada, Ltd. vs Consolidated Mining & Smelting Co. of Canada, Ltd.*, was postponed while this case was heard, since infringement was not an issue if the patent itself was invalid.

46 From the "Reasons for Judgement of the Hon. Mr. Justice Audette," published in the transcript of the case when it was appealed to the Supreme Court of Canada, *Electrolytic Zinc Process Company, Plaintiff vs French's Complex Ore Reduction Company of Canada, Ltd., Defendant*, Vol. 7, 1340. Copy in File 6, Box 194, Add. Mss. 2500, BCARS.

47 *Electrolytic Zinc Process Company, Plaintiff vs French's Complex Ore Reduction Company of Canada, Ltd., Defendant.* Vol. 3, 448, in File 2, Box 194, Add. Mss. 2500, BCARS.

48 *Op. cit.*, 1342.

49 For the judgment of the Supreme Court, see [1930] 4 DLR 902-15. The fact that Fox, in *The Canadian Law and Practice Relating to Letters Patent for Inventions*, refers to the case sixteen times (including a direct quotation from the Supreme Court judgment) suggests that the case has some significance in the history of Canadian patent law.

50 Letter from Vancouver Board of Trade to Premier McBride (also Minister of Mines), 13 November 1915, copy in Vancouver Board of Trade, Minutes of the Board's Mines Committee, 1902–1917, Vol. 127, 51, Add. Mss. 300, City of Vancouver Archives. For ease of reading I have deleted a number of eccentrically placed commas.

51 J. Read, CANADA *Debates*, Vol. 2, 23 May 1918, 2503.

52 See the "BC Ore Dressing Plant" file, No. 160, Vol. 1, in Geological Survey of Canada records, RG45, Vol. 19, NAC.

53 Share data on Cominco, clearly showing a leap in value in 1924–1925, may be found in the relevant copies of the *Annual Financial Review*. It is worth noting that the electrolytic process developed by Cominco (and, virtually simultaneously, by Anaconda in Montana) also had a significant global impact. By 1975, over two thirds of world zinc production came from ore treated by the electrolytic method; see S. W. K. Morgan, *Zinc and Its Alloys and Compounds*, Chichester: Ellis Horwood Ltd., 1985, 20-21.

54 For statistics on Cominco's share of the dividends produced by BC mining companies, see the table compiled by the Department of Mines in 1941, which indicates that up to that date virtually one half of all dividends paid out by BC mining companies (including coal) were paid to Cominco shareholders: "Table XVII. — Dividends Paid By Mining Companies, 1897–1941," *BC Mines Annual Report*, 1941, 18-20. The quotation is from R. J. W. Douglas, ed., *Geology and Economic Minerals of Canada*, Department of Energy, Mines and Resources, Ottawa: 1970, 504. On the relationship between Cominco's prosperity and the Sullivan mine, see Jeremy Mouat, "Creating a New Staple: Capital, Technology and Monopoly in BC's Resource Sector, 1901–1925," *Journal of the Canadian Historical Association*, New Series, Vol. 1 (1990), 215-37.

CHAPTER FIVE

Between the Wars: *West Kootenay Power in Good Times and Bad*

At the end of the First World War, West Kootenay Power's generating plants and transmission lines formed an impressive network by the standards of the time. Yet the company did not stand still. In the two decades between the wars, Lorne Campbell continued to push hydro-electric development along the Kootenay River. First, the original Number One plant was torn down and replaced by a much larger facility. Almost immediately, a new plant was built at South Slocan, followed by a fourth at Corra Linn. A fifth was in the planning stages when the Depression slowed economic activity after 1930.

Cominco, the parent company, realised the critical role that power assumed in its operations and readily supported Campbell's plans. This ongoing industrial growth brought its share of problems and new complexities. For example, the increasing scope of West Kootenay Power's operations meant that its plans had to clear new obstacles. The Kootenay River, site of its core operations, was an international river, crossing the border several times before joining the Columbia at Castlegar. Consequently, when West Kootenay Power wanted to regulate the flow and storage of water along the river and in Kootenay Lake, it needed the support of the Canadian government as well as the permission of the International Joint Commission, the combined American and Canadian group which oversaw cross-border developments. This process proved difficult and time-consuming.

Other cross-border issues drew the International Joint Commission into the region. As Cominco's Trail plant expanded, its growing level of sulfur dioxide emissions became a very contentious issue. The subsequent environmental damage south of the border led to international arbitration, as farmers in Washington state demanded an end to the pollution of their region. Cominco resolved the worst aspects of its sulfur emissions by diversifying its operations even further. The company's engineering staff devised a means to recover much of the sulfur before it was released into the atmosphere and used the material in fertilizer production. Once more, new operations demanded yet more power and pushed West Kootenay Power to extend its operations even further.

A search for new markets for electricity led West Kootenay Power to expand into the booming mining district west of Rossland and Trail. The Boundary region — the operations of BC Copper and Granby Consolidated Mining and Smelting in Greenwood, Phoenix, and Grand Forks in particular — proved at first a profitable field for the company. But mining operations are always limited by the extent of their ore deposits and the activity in the Boundary was no exception to this rule. By the end of the First World War, people realised that declining metal prices combined with dwindling ore deposits spelled the end of the Boundary's once thriving mining industry. The Granby company relocated its operations to the northwest coast, developing a new mine at Anyox. The BC Copper Corporation — re-organized as the Canada Copper Corporation — opted to develop a mine closer to its original property, acquiring the Copper Mountain mine near Princeton. Determined to make the best of the situation, in 1918 West Kootenay Power moved quickly to extend its transmission lines even further westward, to supply power to the new site of Canada Copper's mining operations.[1]

In the autumn of 1920, West Kootenay Power's lines reached Copper Mountain. Construction was slowed due to a labour dispute and the subsequent rush to complete the line resulted in a less than perfect job. This created problems in the mid 1930s when a substantial part of the line collapsed in the middle of a hard winter. Nor did the mine itself have a very successful start. The end of the war brought with it depressed metal prices, and the post-war economic climate was not one ideally suited to launching new enterprises. Less than two months after the transmission lines reached the Copper Mountain mine, operations shut down. Canada Copper could not continue to ship ore and make a profit with copper at thirteen cents a pound. In what amounted to an obituary, the 1920 *Annual Report* of BC's Minister of Mines regretfully noted that "Unless the price of copper advances or the cost of operation

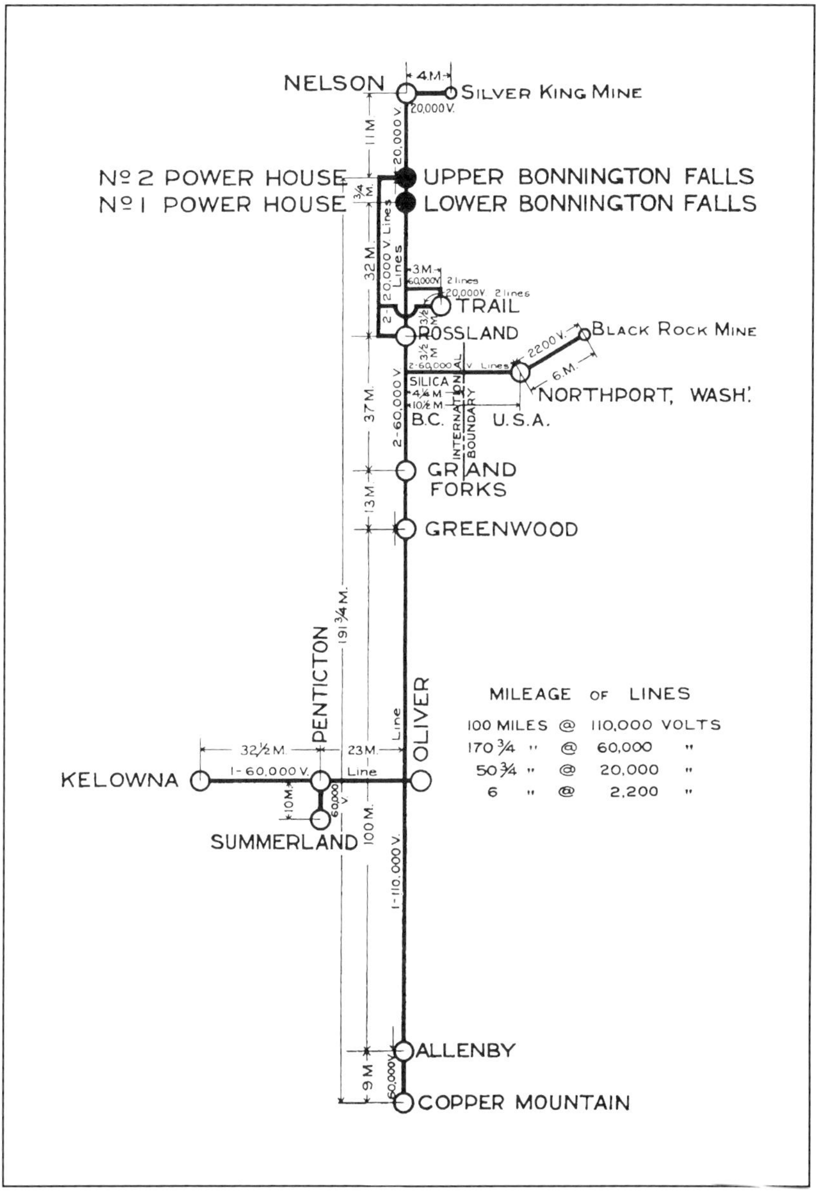

A schematic diagram of West Kootenay Power's transmission lines in 1924. Despite an overall length of nearly two hundred miles, most of the power travelled a relatively short distance, from the two Kootenay River plants to the Trail smelter. The illustration appeared in a booklet published by BC's Department of Lands in 1924, *Water Powers, British Columbia, Canada.*

decreases, it will be some time before the mine and mill commence to operate."[2] Undaunted by the setback, West Kootenay Power continued to search for opportunities in the South Okanagan.

Efforts to irrigate the region, thereby making it suitable for extensive agricultural operations, required power to drive large pumps. The government-led Southern Okanagan Irrigation Project, which would bring water to the Oliver and Osoyoos area and provide farming opportunities for First World War veterans, became one outlet for the electricity available from West Kootenay Power's new lines into the region.[3] The towns of the region, stretching north to Kelowna, were another possible market. A municipal steam plant had provided Kelowna with electricity from 1908 but by the end of the 1910s this power source was becoming unequal to the task. In early 1922, city residents happily endorsed the arrangement that the municipal authorities negotiated with Lorne Campbell. Before the year's end, their homes enjoyed the benefits of West Kootenay Power.[4] The arrangement was typical of those concluded with other towns in the region: West Kootenay Power provided the municipal authority with electricity, which in turn distributed it to local residents.

Although the company was likely grateful for their custom, neither the towns of the region nor its irrigation pumps called for a considerable amount of electricity. The Okanagan did not become a profitable part of West Kootenay Power's network until after the Second World War. What drove the company's expansion between the wars was the rapid growth of Cominco's operations and the subsequent need for an ever-increasing power supply.

The last chapter described Cominco's success with zinc treatment which permitted it to exploit the full potential of the Sullivan mine's ore. The company's strides in metallurgy and its subsequent growth were widely recognised. For example, in 1925 the *Annual Report of the Minister of Mines* devoted considerable space to Cominco's increasingly prosperous operations. Seven pages provided a lengthy description of the Trail plant, concluding that the smelter now ranked "high among the greatest metallurgical plants of the world." The *Report*'s frontispiece was a panoramic photograph of Trail, graphic evidence of Cominco's predominance. As we'll see, the billowing smoke from the smelter's stacks — intended, no doubt, to show an industrial complex in full production — was to create considerable controversy in the years ahead.

The *Report*'s description of the Trail smelter noted that the electrolytic zinc plant consumed more than 70% of Cominco's power requirements.[5] This meant that the rapid expansion of zinc production at Trail

Among the greatest metallurgical plants of the world . . .

During the years following the successful development of the Sullivan mine it [the Trail plant] has undergone extensive alterations and additions, until now it ranks high among the greatest metallurgical plants of the world. . . . The urgent need for zinc by Great Britain and her allies during the Great War was largely responsible for energetic and aggressive action on the part of the company to produce this metal, and with this object in view the erection of an electrolytic zinc-refinery was commenced and rushed to completion in 1915 and put into operation in February, 1916. This marked the beginning of a new era for the company, which then, naturally, had not only to turn its attention to solving the problem of recovering the values in the complex ores of the now world-famous Sullivan mine, but also had to face the many problems in connection with the various phases of the electrolytic refinement of zinc, which was at this time in the experimental stage. . . .

– *Annual Report of the Minister of Mines for 1925*

had to be matched by a similar increase in West Kootenay Power's generating capacity. And by the 1920s, the modest volume of power produced by the old Number One plant was no longer equal to this challenge. The time had come for a much more ambitious scheme.

As a direct result of Cominco's expanding zinc production, a new facility was designed and built on the site of the old Number One plant at Lower Bonnington. Demolition of the original brick building began in 1923 and in the summer of 1925 a much larger concrete building stood in its place. The old Number One plant housed three generators, capable of producing a total of 4,000 horsepower. Its replacement boasted three 20,000 horsepower turbines, operating with twice the head — that is, double the vertical distance from the water intake at the headworks to where the water was discharged below the dam — of the original plant.[6] The fifteen-fold increase in generating capacity was dramatic evidence of the company's growth, but equally significant was the fact that this was truly a West Kootenay Power plant. Unlike the original Number One and the Number Two plants, the engineering design was done by company staff.[7] This would also be the case with the later plants built along the river

Lorne Campbell was no doubt proud of the new plant and he decided to celebrate its opening with an official ceremony. On the first of August 1925, over five hundred people were on hand to see T. D. Pattullo — then Minister of Lands and later premier — throw the switch to start the plant's generators. Fifty years later, employees of West Kootenay Power who had attended the opening recalled the event with some amusement.[8]

> They had a fairly wild party! . . . the Superintendent had had quite a few. . . . He came down and I could see the condition he was in. He took [Pattullo] over to the switchboard and told him to pull the button. It was the right button, but knowing that L. A. Campbell was down there too I was watching very carefully. I saw the voltage mounting up and up until it reached the proper voltage and then it kept going up. I walked over to [Pattullo] and said, "Would you please release the button as we are a little too high and we are not ready to pull that switch yet." He let go of the button and I gently pushed it back until we got to the voltage we needed. . . .
>
> — C. H. BLAND, Interview #21, 19 August 1975, 2

> I remember they were trying to synchronize. We knew why they didn't get it synchronized. They were so full of John Barleycorn that night that their noses were out of joint about pulling the switch to get the generator back on, and after [Pattullo] had pulled it on why this fellow . . . turned around and reached across the board and pulled it off again and then that kind of upset the apple cart. . . . The big boss L. A. Campbell . . . he got the superintendent off to one side and he fired him right there, just like that. There was a big dance going on up in the hall, big crowd up there, the superintendent comes up there pretty well lit and at the top of his voice he roars, "The old son-of-a-bitch fired me." Pardon the language. He hired him back on later. . . .
>
> — DAVIE MCDONALD, Interview #4, 16 June 1975, 6-7

> Talking about the opening of No. 1. . . . Jess Barwise, he had the unfortunate job of being elected to operate a search light. They put up this big search light on the roof, projected onto the falls, it was quite spectacular in the night. While this party was going on, everyone would say, "Poor old Jess, he's stuck out on that roof all night and he hasn't got a thing to drink." So I think, "By golly, I'd better take Jess out some beer," so I gathered up about four bottles of beer and took it out there. Poor old Jess, he used to hum a little song to himself sort of under his breath, he was quite happy with this big search light. "Do you care for a beer, Jess?" "Sure, sure, put it in the box." . . . He had three dozen there already. . . . Then the coast train came through. It was just nicely dark and I saw the headlights [of the train] coming down. Jess says, "Who's this coming, I've got competition here," and he swings [the search light] around right on the locomotive as it came around the corner. The poor engineer, all he could see was apparently another locomotive on the same track coming right towards him. He slammed on the emergency, you could hear the wheels a-grinding there for five minutes. I said, "Jess, for God's sake, get that thing off the locomotive, you stopped the passenger train." "Oh, to

hell with him," he turns it back again and the engineer says "toot, toot," and took the air off. Never heard a thing about it. . . .

— Interview #4, 16 June 1975, 5

More formal records of the event also survive. The company went to the trouble of producing a commemorative pamphlet to mark the opening and the *Annual Report of the Minister of Mines* in 1925 included a fulsome paragraph, drawing its readers' attention to the significance of the occasion.[9]

This splendid installation . . .

An event that marks the progress of mining in a most realistic way was the official opening of the new power plant of the West Kootenay Power and Light Company at Bonnington on August 1st by the Hon. T. D. Pattullo, Minister of Lands. . . . This splendid installation, representing the latest word in hydro-electric engineering, reflects great credit on Lorne A. Campbell, who has successfully guided the destiny of this company for so many years. The power derived is mostly consumed by the Consolidated Mining and Smelting Company at its great metallurgical works at Trail, the increased capacity of which has demanded more power than hitherto available.

– Annual Report of the Minister of Mines for 1925

As the celebration of the official opening suggests, the completion of the new Number One plant was a milestone of sorts for West Kootenay Power. The company could now design and build its own dams and generating plants, and would be able to expand as Cominco continued to increase the size of its operations. In fact, this happened almost immediately. Cominco carried out extensions to its zinc plant — the most significant energy consumer in the company's Trail operations — in 1925, in 1927, and again in 1929.[10] "Rapidly expanding zinc production absorbed almost all of the power from the new No. 1 Plant," recalled an article in the *Canadian Mining Journal*, "and in the year following its completion a start was made on a new power project."[11]

South Slocan — the name given to the Number Three plant — was begun in September 1926.[12] It was designed to function as West Kootenay Power's centre of operations, to be the main control and switching station for all the company's plants along the Kootenay River. The plant's construction was the largest project yet attempted by the company. Eventually it would have three turbines producing a total of 75,000 horse power. But the first step was to build facilities to house and feed the six hundred construction labourers that would be needed to carry out the work. As well as the inevitable bunk-houses and eating areas, the company built a recreation hall which accommodated a bowling alley and movie projector; films were shown twice a week. Other popular leisure activities included weekend dances, badminton, and various card games (frequently accompanied by gambling). A long-time employee recalled many years later with a laugh that "they did a lot of drinking, fishing and hunting and fighting."[13] The work itself was certainly challenging, involving the excavation of some 160,000 cubic yards of rock and pouring 120,000 cubic yards of concrete. Much of this work was done by hand or with the help of what would be regarded by today's standards as rather crude mechanical tools. But the biggest contrast between then and now was the pay. Labourers made a little over three dollars a day (forty cents an hour), while the highest paid tradesmen earned seventy-five cents an hour.[14] When the Depression came, wages sank even lower.[15]

People who worked on the company's construction projects often had mixed memories of the period. Many of the employees on construction were unlike those who carried out the more routine tasks of regular operation and maintenance:

> On construction it is a little bit more free and easy. . . . Actually, on construction there is a different type of person . . . they are boomers, so to speak. They go from one construction job to another and the guys don't have too much respect for authority, you know. It doesn't take them much to tell the superintendent or the foreman what they can do, and [then they] go and pull their time and go someplace else.[16]

Almost everyone had their favourite stories from the period, accidents that in retrospect seem amusing rather than disastrous. R. J. Mulloy remembered one incident involving the "dinkies"—small gas-driven locomotives that ran on tracks and were used to haul the rock out of the glory hole, once the rock had been blasted loose. Mechanical shovels would load the rock onto cars and then the dinkies would haul the cars up out of the glory hole so that the rock could be dumped out of the way. It was rough work, hard on the equipment, and regular maintenance was essential if operations were to keep going without any breakdowns.

> . . . this is rock they are loading, very hard on the cars . . . the rock was dumped by the shovels [and] the cars would get damaged, so we had a shed up there where we had carpenters [who would] repair the cars, we also had a section there for . . . the dinkies, you see. . . . The idea was we

would bring them out of the hole when their shift was ended and there was going to be a gap in there, [a period when] we would try to do their maintenance. You don't tie things up on construction, you fit yourself in. . . . So the idea was there would be maybe one or two men down in the glory hole, in the area where they were excavating the rock. They started the dinky up, you see, and they started up the hill. They would be on the tracks and they would leave [the dinky] in gear and they would just go through the switches which was all set and . . . then a man up on the top would catch the dinky and he would climb on board as it goes by and kick it out of gear, and spot it, and wait for the next one to bring it in carefully and line them up and that. . . .

Jimmy was up top just about to catch these dinkies and the phone rang . . . so up the hill goes Jimmy. Jimmy just gets in there to answer the phone and then a look of horror comes across his face: he realizes that the dinkies are coming up. He just dropped the phone and out the door, just in time . . . to see a dinky disappear into the dinky shed going full bore, come out the other end with the pot-bellied stove on the front of it and the work bench and everything else, you see. Down the hill he goes [but] he didn't make it for the second one. The second dinky went in and, well, it torpedoed the chaos. It took part of the bench and here was two dinkies derailed . . . no end in the dinky shed, the stove was out there, tools, benches and everything else.[17]

But the same man could remember darker moments as well:

Safety was the name of a new soft drink; it was never heard of, it was nothing. . . . Ask any of the old-timers. . . . I told you about the glory hole there, they had these kids down in the bottom there, loading these skips. . . . There were a bunch of young fellows, you see, hanging in bosun's chairs, scaling down the loose rock with bars. . . . If these kids down below moved back while they were scaling [rock above them], they were liable to get fired, see. They had to keep on with the rocks running all around them. Oh, they just treated the kids like dirt. Then suddenly the war started [and] instantly they weren't a bunch of bums, they were our dear boys going to fight and defend our country, they were recruited. Do you know what our dear boys did? . . . They laid hands on every bit of equipment they could—jack hammers, jacks, everything they could lay their hands on—and dumped it in the river and drew their time and went in the army. . . . These were all young fellows, as I say, 18, 19, maybe 20, darn good kids too, but they were on a bull gang run by a great big hulking raw-boned Scotsman who didn't treat them very nicely at all. . . . But that was the situation. That's the way it was on construction, when I lost this finger.[18]

The South Slocan plant was fully operational by the summer of 1929, but more construction was soon being planned. The Corra Linn dam was the next project to be built, higher up the river and closer to Nelson than the other plants. However, the story is a little more complicated than that, since West Kootenay Power had earlier contemplated building a dam at Granite, just three miles downstream from Nelson.[19] In September 1929 the company formally requested permission from the International Joint Commission (IJC) to build the Granite Dam. The IJC — made up of both Canadian and American representatives — was formed just before the First World War to adjudicate cross-border issues, particularly those involving shared boundary waters.[20]

International Considerations

The Boundary Waters Treaty of 1909 was designed to insure the constructive utilization in the best and in the mutual interests of the High Contracting Parties [Canada and the US] of boundary waters and waters flowing across the boundary. The Treaty set up in the International Joint Commission the necessary machinery to realize this objective.

The Kootenay River is probably the most outstanding example of an international stream along the entire boundary. . . . In order that the United States and Canada may realize the development of the full resources of this international river system, it is essential that there should be the fullest possible measure of mutual co-operation at all points. Any other policy can only prove mutually hurtful. . . . There are many international features in relation to this watershed which will require the adjudication of this Commission, and, in the consideration of this present matter, I think it is very desirable to keep the general picture before our eyes and also to keep in mind the fact that mutual co-operation is necessary to secure the utmost mutual use of the resources of the watershed.

— **J. T. JOHNSTON** before the International Joint Commission, 1933

The purpose of the Granite Dam was not simply to generate power but also to control the flow of the Kootenay River. West Kootenay Power needed the IJC's approval since the dam would in effect act like a tap: at times during the year it would hold water back in Kootenay Lake, conserving it so that it could be used during periods of what otherwise would have been low water. When the water was being held in the lake, however, it could flood land higher up the river, in the area around Creston, BC, as well as in Idaho. The possible impact of the dam on northern Idaho explains the involvement of the IJC.

Once the IJC received the company's application to build the Granite Dam and to store up to six feet of water in Kootenay Lake, it held hearings at Bonners Ferry, Idaho. Farmers in the area as well as state

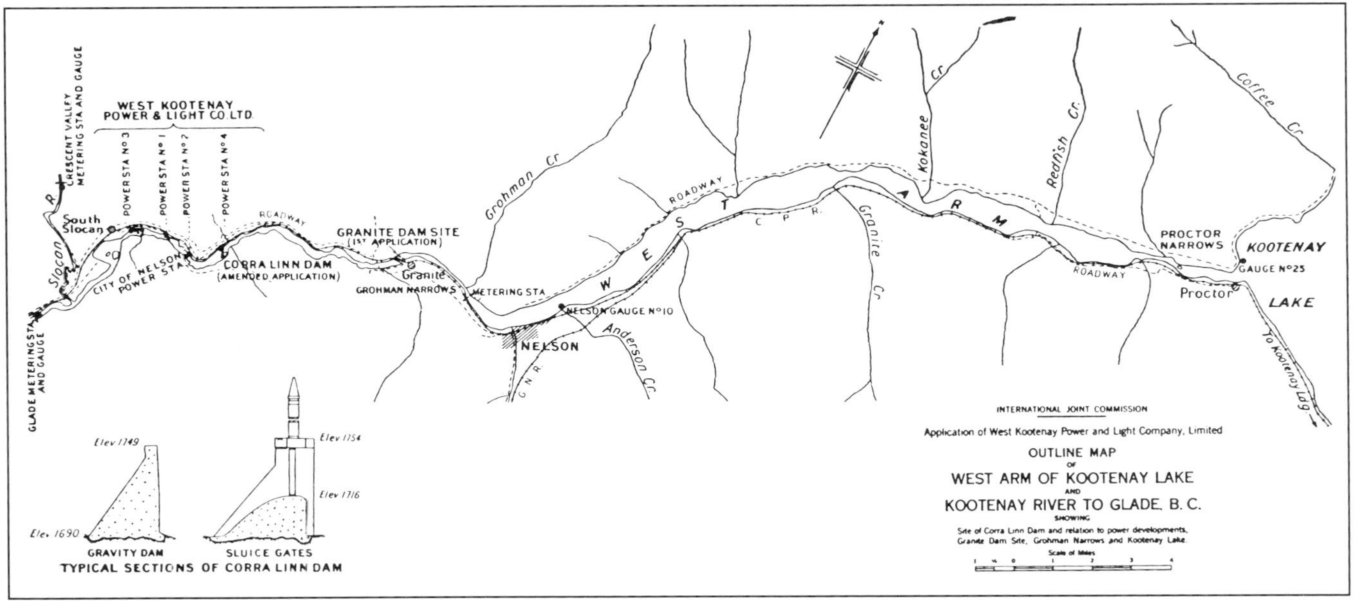

This "Outline Map of West Arm of Kootenay Lake" accompanied West Kootenay Power's application to the International Joint Commission to raise the level of Kootenay Lake. [Printed in *The Kootenay Valley: A report on certain cases involving reclamation and the development of water power in the valley of the Kootenay River, under the terms of article IV of the Treaty of January 11, 1909*, Ottawa: J. O. Patenaude, King's Printer, 1936, 127]

The Trail smelter in full production.

officials objected to the construction of the dam. In their view, prior to approving the project, the IJC should insist on extensive studies so that it could carefully assess the dam's impact.[21] But West Kootenay Power was in a hurry. When the Bonners Ferry hearings made it clear that the construction of the dam would inevitably be delayed, the company reconsidered its options. It abandoned the original plan for the Granite Dam and instead went ahead with a similar project a little further downstream, the Corra Linn Dam. Although the Corra Linn would potentially affect water levels in much the same way as the Granite Dam, this would only happen if the company chose to do so. That eventuality could wait for the IJC's approval; in the meantime, construction could proceed.[22]

West Kootenay Power needed to rush construction of a fourth plant for two closely-related reasons. First, the company wanted to ensure that all its plants along the Kootenay River ran at full capacity, and to do this it needed to stabilize the flow of water. For example, the new South Slocan plant needed 10,500 cubic feet of water per second to run at full capacity. However, during the winter the river flow could occasionally drop to less than half this amount. To overcome this problem, the company's engineers wanted a dam at the lake's outlet west of Nelson. Such a dam would enable them to use Kootenay Lake as a temporary reservoir to hold some of the fall run-off, to be released later during the dry winter months. The second reason for the dam was more straightforward: the company urgently needed to be able to produce more power. Cominco was in the process of installing a new plant and this would require a substantial amount of energy. Ironically, this reflected another issue that concerned the International Joint Commission.

Cominco's hectic pace of expansion during the 1920s had created some new problems for the company. The most pressing of these was the environmental damage caused by the growing amount of sulfur dioxide going up the smelter stack in Trail.

A Cominco Engineer Explains . . .

. . . a zinc plant was started in 1916, in response to the demands of wartime. The amount of sulfur in the ores treated at Trail had by now risen to about 140 tons per day, and about 90% of it went up the stack as SO_2, mostly at a concentration of about 1 to 2%. There was definitely quite a lot of localized damage to vegetation, but this was a rather remote mountainous area; most of the inhabitants made their living at the smelter; the smelter had acquired smoke-damage easements on most of the habitable land; and at the time no one was particularly disturbed about the situation.

Meanwhile, research work was started on the ore from the Sullivan mine, and a separation process was developed utilizing the recently discovered idea of froth flotation. A flotation mill was brought into operation in 1923, and by 1926 the metal output, and with it the sulfur output, was more than doubled. It was shortly after this that Cominco started to run into trouble with atmospheric pollution.

— J. N. ROBINSON, Chemical and Fertilizer Operations, Cominco, Ltd., 1972

At first, Cominco tried simply to disperse the smelter emissions more efficiently.[23] The company built two four-hundred-foot chimneys, at the time a common way to deal with the problem of airborne pollutants. However, local atmospheric conditions and the topography of the area combined to make this solution largely ineffectual. Often the smelter smoke would be held within the narrow confines of the valley, causing considerable damage to vegetation. The forty-ninth parallel was very close—just over ten miles down the valley—and Washington state farmers soon began to protest the situation.

The strategy that Cominco had tried around Trail—purchasing land affected by pollution or negotiating some other way to silence those who protested of the pollution—could not work in Washington.[24] State law did not allow a foreign company to own or control land in Washington. On the other hand, aggrieved American farmers could not bring Cominco to court in the United States, since the courts did not have extra-territorial jurisdiction. However, in 1928 the American government — prompted by Washington state — referred the matter to the International Joint Commission.[25] The US government identified five issues for the IJC's consideration: the extent of damage that the smelter fumes had caused south of the border; the level of compensation that would adequately reimburse those who had suffered this damage; the impact in Washington state of continuing smelter operations; a method of compensation for ongoing damage; and any other issues that might arise from this specific instance of cross-border pollution. The Canadian government agreed that the matter should go before the IJC, and hearings were subsequently held in both British Columbia and Washington state.

The International Joint Commission's report on the case appeared in early 1931. This recommended that $350,000 be paid to cover the damages suffered up to the end of 1931. (Cominco would pay this amount to Ottawa and then the federal government would forward the money to the US government.) The American government did not accept the report, however. Farmers in Washington state remained unhappy, arguing that their properties continued to suffer damage as a result of the smelter's emissions. Negotiations between the US and Canadian governments led in the spring of 1935 to the creation of a three-person arbitration tribunal, made up of an American representative, a Canadian, and a third person acceptable to both parties.[26] The tribunal was to assess the extent of any further damage which the smelter may have caused since the IJC report, and also whether the smelter's operations ought to be curtailed or altered so that damage would no longer occur in Washington state.

The arbitration tribunal commissioned a series of exhaustive technical reports by scientists and conducted what was in effect a two year experiment on the impact of sulfur dioxide in the region.[27] As a result of this work, it ruled that "the operation of the Smelter and the maximum emission of sulphur dioxide from its stacks shall be regulated" and carefully detailed a regime for the smelter to follow.[28] This was plainly intended to minimize damage, particularly to crops during the growing season. The tribunal also determined that a further $78,000 damages had occurred between the end of 1931 and 1937, in addition to the $350,000 damage assessed by the IJC. In its final decision of 1941, however, the tribunal concluded that no damage had occurred after 1937.

ROSSLAND HISTORICAL MUSEUM

West Kootenay Power's new office block in Trail opened in 1930. Above, the building is adorned in patriotic fashion to celebrate George VI's coronation in 1937, while refrigerators crowd the showroom windows at ground level. Below, visitors to the building encountered a wide array of electrical appliances. Hidden lights cast beams up to the ceiling, illuminating the plaster representations of electrical generators.

ROSSLAND HISTORICAL MUSEUM

The tribunals's conclusion that damage had significantly declined by the late 1930s reflected changed conditions at Cominco. Company staff devised a way to recover sulfur dioxide from the flue gas prior to its release into the atmosphere. As a consequence, the smoke's toxicity was considerably reduced. (The tribunal noted, for example, that by the late 1930s the smelter was releasing well under half the amount of sulfur dioxide that had been going up the stacks in 1930.[29]) And Cominco made a virtue of necessity, for the recovered sulfur became the basis for a new and profitable sphere of operations, fertilizer production.[30] These developments had a considerable impact on West Kootenay Power's operations, for Cominco's new program of sulfur recovery called for a substantial amount of electrical energy.

West Kootenay Power made no secret of the reason why it submitted its application to the IJC in the autumn of 1929. The company's lawyer explained that the Granite Dam was necessary so that Cominco could have sufficient power to press ahead with its pollution control program.

> [Cominco] is going to be called upon within the next two years to get into operation a large plant for the manufacture of fertilizer, for the operation of which a large quantity of power will have to be supplied within the next two years, [and so West Kootenay Power] finds it necessary to supply such power within the said time and has no means at its disposal to do so except through the additional power that will be obtained from this storage. At the present time during the winter months there is insufficient power for the industries and communities now consuming power, and [Cominco], being the chief customer of the applicant, is required to curtail its operations during the said period. With the new fertilizer plant in operation, the situation will be even more acute unless further power can be generated. This additional power required for the fertilizer plants of [Cominco] being required in consequence of the efforts of [Cominco] to mitigate another international problem, it seems reasonable to me to be asking this Honourable Commission to facilitate as much as possible the final disposition of this matter.[31]

The company reiterated this argument when it re-applied for permission to raise the level of Kootenay Lake in 1932, once the Corra Linn Dam had been built.[32] Technically this second application to the IJC was simply an amendment to the company's earlier one involving the Granite Dam, since the IJC had adjourned in 1929 without making a decision. In fact, much the same thing happened in 1932 as with the company's original application: hearings were held, objections were put forward, and the company finally withdrew its application. But the IJC did eventually allow the company to raise the level of Kootenay

Lake, when it applied yet again, in the autumn of 1938. Ironically, the people responsible for this third application were the Idaho farmers who had earlier opposed the dam's role as a control device on Kootenay Lake. Heavy winter flooding in the 1930s had taught them the virtues of flood control, and they petitioned West Kootenay Power to re-apply to the IJC. On 11 November 1938, the IJC approved the company's plans to store water in Kootenay Lake.[33]

West Kootenay Power became a large and sophisticated utility during the 1920s and the 1930s. Three large construction projects—the new Number One plant, South Slocan, and Corra Linn—had greatly increased the company's generating capacity. When the First World War ended in 1918, it could produce 38,000 horse power; when the Second World War began in 1939, this figure had multiplied by a factor of six, to 226,000 horse power.[34]

If West Kootenay Power remained largely an industrial supplier with its career intimately linked to that of its parent company, Cominco, the future seemed promising. Some employees confidently imagined an era of continuous expansion and growth. The following comments from a retired worker suggest the optimism of the years between the wars:

> When we first started on construction at the Corra Linn plant, we thought we were going to go from the Corra Linn construction to the Brilliant plant construction and then on to construct plants on the sites on the Pend d'Oreille River. We thought the company was really going to expand and be in a position where they could take on a lot more of the utility load in the Interior of the province. Because of the Depression, when they cut back on their program, they did not go ahead with the Brilliant power plant until after the war and it was then a different type of arrangement entirely. West Kootenay never did develop the sites on the Pend d'Oreille River. We were a little disappointed as we thought at that time that we were going to have a real up-and-coming energetic program of expansion in the public utility field. We used to compare West Kootenay with BC Electric, which was the utility in the Lower Mainland. I think we were probably dreaming but we had an idea that West Kootenay would take over the whole of the Interior of the province, everything east of the Coast Range, and BC Electric would have the coast. . . . The Depression set in and, of course, they cut back on their program and they never did build the other plants and only supplied power as required by Cominco from then on. . . . When the BC Power Commission was formed by the government, and then eventually they took over BC

Electric and formed BC Hydro and expanded right into the Interior of the province, further expansion by West Kootenay was cut and stopped right at Princeton.[35]

As these comments hint, much was to change in the years after 1938. Up until that time, public utilities in British Columbia had hardly felt the hand of the government regulator. All that changed in December 1938 with the passage of "An Act to provide for the Regulation of Public Utilities." The Second World War also brought important changes: the world after 1945 was in many ways a very different place.

NOTES

1 For detail of Granby's shift to Anyox, see Barnard Powers, "Mining and Oil: What Happened to Granby," *Magazine of Wall Street*, Vol. 17, No. 5 (11 December 1915): 337-39; C. M. Campbell, "The Granby Mines at Phoenix," *Transactions*, Canadian Institute of Mining and Metallurgy, Vol. 22 (1919): 155-79; and L. E. Carter's "Granby: Seventy-five Years of British Columbia Copper Mining," BA Hons. Essay (Geography), Simon Fraser University, 1979. On the Canada Copper Corporation and West Kootenay Power's line construction, see the accounts in British Columbia, *Annual Report of the Minister of Mines for 1918*, K201; *Annual Report of the Minister of Mines for 1919*, N162; and *Annual Report of the Minister of Mines for 1920*, N159.

2 British Columbia, *Annual Report of the Minister of Mines for 1920*, N159.

3 See the bound volume — containing photographs and details of the scheme — in British Columbia, Water Rights Branch, Southern Okanagan Irrigation Project, GR1288, BC Archives and Records Service. Further information on the scheme may be found in Robin Fisher, *Duff Pattullo of British Columbia*, Toronto: University of Toronto Press, 1991, 141-45, and Paul M. Koroscil, "Soldiers, Settlement and Development in British Columbia, 1915–1930," *BC Studies*, #54 (Summer 1982): 63-87.

4 See "Power Company Contract Explained," *Kelowna Courier and Okanagan Orchardist*, 29 December 1921, 1, 8; "Thumping Majority For Power By-Law," *Kelowna Courier and Okanagan Orchardist*, 5 January 1922, 1; and "From Steam to Hydro-Electric on Sunday Next," *Kelowna Courier and Okanagan Orchardist*, 26 October 1922, 1, 4.

5 "The power requirement for the zinc plant is 28,000 kw. and for other parts of the Trail plant is 11,000 kw., making a total of 39,000 kw. or 52,200 horse-power. This power is developed at Bonnington Falls by the West Kootenay Power and Light Company, a subsidiary company of the Consolidated Mining and Smelting Company." (British Columbia, *Annual Report of the Minister of Mines for 1925*, A257).

6 The new plant had seventy feet of head, compared to the thirty-four feet head of the original plant. This increased head was a result of lowering the level of the river below the dam: "The additional fall was gained principally by the removal of an obstacle in the river below the plant and the lowering of the level of the river." (F. H. Mason, "Bonnington Falls Hydro-Electric Station," *Canadian Mining Journal*, Vol. 50 [25 January 1929], 72).

7 An article describing the new facility noted that "The plant was designed by Lorne A. Campbell, vice-president and general manager of the West Kootenay Power & Light

Co., assisted by E. M. Stiles, W. Tindale, J. D. McDonald and A. E. Wright" (Lorne A. Campbell, "Development at Bonnington Falls, BC," *Canadian Engineer*, Vol. 50, No. 3 (19 January 1926), 144). The Number One Plant, wrote the *Canadian Mining Journal* thirty years later, "was the first power house to be constructed after Cominco acquired the controlling interest in the West Kootenay Power and Light Company, and the general arrangement was so satisfactory that it served as a basic model for all subsequent developments on the Kootenay River" ("The Cominco Story: Power," *Canadian Mining Journal*, Vol. 75, No. 5 (May 1954), 331).

8 The following comments are taken from transcriptions of twenty-seven interviews with West Kootenay Power employees. The interviews were tape recorded by Jamie Cameron in 1975. Copies of the transcriptions are held in the West Kootenay Power office in Trail. The original tapes are evidently in the attic of the company's warehouse at South Slocan.

9 A copy of the pamphlet may be found in the provincial archives in Victoria, call #NWp971.91 W5220. The reference in the *Annual Report of the Minister of Mines for 1925* occurs on A249. See also "West Kootenay Power Company Opens Mammoth New Hydro-Electric Plant," *Nelson Daily News*, 3 August 1925, 1, 3, 8, as well as the technical description by Lorne A. Campbell, "Development at Bonnington Falls, BC," *Canadian Engineer*, Vol. 50, No. 3 (19 January 1926): 141-44.

10 A. G. Dickinson, "Utilization of Electric Power at the Metal and Fertilizer Plants, Trail, BC," *Engineering Journal*, Vol. 20, No. 9 (September 1937), 721.

11 "The Cominco Story: Power," 331.

12 The following account is drawn from the descriptions in "West Kootenay Completes No. 3 Plant at South Slocan, BC," *Contract Record and Engineering Review*, Vol. 43 (16 October 1929): 1242-49, and "Power Development at South Slocan, BC," *Canadian Engineer*, Vol. 59, No. 3 (15 July 1930): 141-45. (Note, however, that the second article draws heavily on the first.) See also the excellent overview by long-time West Kootenay Power engineer, W. J. Tindale: "Power Development on the Kootenay River," *Engineering Journal*, Vol. 31, #5 (May 1948): 283-89.

13 Interview #11 (Dave Bell), 7. Mr. Bell was hired in June 1931 to work on the Corra Linn project.

14 Interview #6 (R. J. Mulloy), 7; Interview #11 (Dave Bell and Grant Hall), 6; Interview #18 (Grant Tindale), 6.

15 Interview #10 (Gerry G. Service), 7; Interview #14 (Art Lambert), 3; Interview #23 (J. E. Parker), 6-7.

16 Interview #6 (R. J. Mulloy), 5.

17 Interview #6 (R. J. Mulloy), 3-4.

18 Interview #6 (R. J. Mulloy), 6-7.

19 The following account rests largely on the voluminous report — nearly four hundred pages long — published by the International Joint Commission: *The Kootenay Valley: A report on certain cases involving reclamation and the development of water power in the valley of the Kootenay River, under the terms of article IV of the Treaty of January 11, 1909*, Ottawa: J. O. Patenaude, King's Printer, 1936. For a brief summary, see L. M. Bloomfield and Gerald F. Fitzgerald, *Boundary Waters Problems of Canada and the United States (The International Joint Commission 1912–1958)*, Toronto: The Carswell Company, Ltd., 1958, 125-28, 130-33.

20 For the IJC's role and history, see Bloomfield and Fitzgerald, *Boundary Waters Problems of Canada and the United States,* and Robert Spencer, John Kirton, Kim Richard Nossal, eds., *The International Joint Commission Seventy Years On,* Toronto: Centre for International Studies, University of Toronto, 1981.

21 See the newspaper account of the hearings: "Idaho Bucks Nelson Dam," *Nelson Daily News,* 7 November 1929, 1; also IJC, *The Kootenay Valley,* 106-107, 111, 121-23, 300-304.

22 See the comments of West Kootenay Power's lawyer, in IJC, *The Kootenay Valley,* 134.

23 The comments in this paragraph are largely drawn from J. N. Robinson, "The History of Sulfur Dioxide Emission Control at Cominco Ltd.," *International Journal of Sulfur Chemistry,* Part B, Vol. 7, No. 1 (1972), 51-53.

24 For Cominco's earlier efforts in and around Trail to deal with those who complained of the smelter's pollution, see James Robert Allum, "Smoke across the Border: The Environmental Politics of the Trail Smelter Investigation," PhD thesis (History), Queen's University, Kingston, Ontario, 1995, esp. 99-130.

25 The case is summarised in Bloomfield and Fitzgerald, *Boundary Waters Problems of Canada and the United States,* 137-38, and discussed more fully in "Trail Smelter Case," United Nations, *Reports of International Arbitral Awards/Recueil des Sentences Arbitrales,* Vol. III (1949): 1905-1982, 2055-56. For closer analysis, see John E. Read, "The Trail Smelter Dispute," *Canadian Yearbook of International Law/Annuaire Canadien de Droit International,* 1 (1963): 213-29; Alfred P. Rubin, "Pollution by Analogy: The Trail Smelter Arbitration," *Oregon Law Review,* Vol. 50 (1971): 259-98; D. H. Dinwoodie, "The Politics of International Pollution Control: The Trail Smelter Case," *International Journal,* Vol. 27, No. 2 (Spring 1972): 219-35; and Keith A. Murray, "The Trail Smelter Case: International Air Pollution in the Columbia Valley," *BC Studies,* 15 (Autumn 1972): 68-85. For an extensive—and highly critical—re-interpretation, see Allum, "Smoke across the Border," *passim.*

26 See the description of these events offered by the prime minister: Canada, House of Commons, *Debates,* 1935, Vol. 4 (10 June 1935), 3455. The tribunal's judgment—published in a sixty-one-page pamphlet—includes an account of its formation; see *Trail Smelter Arbitration between the United States and Canada under Convention of April 15, 1935: Decision of the Tribunal Reported March 11, 1941,* Washington: US Govt. Printing Office, 1941.

27 See *Trail Smelter Arbitration,* 37-47.

28 *Trail Smelter Arbitration,* 47.

29 *Trail Smelter Arbitration,* 12. A graph showing the "Relation of Sulphur Recovered to Total Sulphur in Charge at Trail Operations" (covering the years 1901–1952) indicates the company's efforts to reduce its pollution; the graph may be found in "The Cominco Story: The Smoke Control Department," 310.

30 For the company's fertilizer production, see J. N. Robinson, "The History of Sulfur Dioxide Emission Control at Cominco Ltd.," *International Journal of Sulfur Chemistry,* Part B, Vol. 7, No. 1 (1972): 51-56, and "The Cominco Story: Chemicals and Fertilizers Division," 287-311.

31 R. C. Crowe's statement before the IJC at the Bonners Ferry hearing, 6 November 1929, in IJC, *The Kootenay Valley,* 113. (Crowe appeared before the IJC in his capacity as the company's lawyer.) There seems little reason to doubt this claim. In 1937, for example, fully 70% of the electricity that West Kootenay Power generated on the Kootenay River

went to Cominco. The Trail zinc plant needed nearly half of Cominco's power requirements, and the new chemical and fertilizer branch in nearby Warfield consumed another 40%. (These figures are given in Dickinson, "Utilization of Electric Power," 719.)

32 See IJC, *The Kootenay Valley*, 134, 136.

33 See Bloomfield and Fitzgerald, *Boundary Waters Problems of Canada and the United States*, 130-31.

34 These figures are from Tindale, "Power Development on the Kootenay River." As he points out, in March 1940 the company was generating 276,000 hp along the river, when two new turbines — 25,000 hp each — were added to the Number Two plant (288).

35 Interview #23, J. E. Parker, 10.

CHAPTER SIX

Another War: *The Second World War and its Legacy*

West Kootenay Power's growth through the 1920s and the 1930s—spurred largely by Cominco's continuous expansion and need for more and more power—had consequences that no one could have foreseen. Shortly before the Second World War, physicists had speculated on the theoretical possibility of atomic explosions. Once the war began, theory and public debate gave way to frantic research, as scientists tried to develop a practical way to build an atomic weapon. The American government, in particular, devoted considerable resources to its atomic program. Cominco's chemical and fertilizer plant at Warfield, amply supplied with West Kootenay Power's electricity, was North America's leading producer of electrolytic hydrogen. This attracted the attention of American scientists since they knew that it was possible to make heavy water—a critical ingredient for their experiments—from electrolytic hydrogen. The world had changed dramatically since Sir Charles Ross and his partners first envisaged a power plant on the Kootenay River.

Not all the changes were dramatic. Wartime also brought with it a much greater degree of government regulation for Canadians. Most people accepted that the increased level of government intervention and control was a necessary consequence of the war. This new government role led to important changes at both the provincial and federal level. One consequence was far more public scrutiny and accountability for utility companies such as West Kootenay Power. If utility regulation came late to British Columbia—almost every other jurisdiction in Canada had had some form of utility regulation for years—its arrival marked the beginning of a series of significant changes in the relationship between Cominco and West Kootenay Power.

Given the strategic importance of Cominco's operations — amply demonstrated during the First World War — it was scarcely surprising that the outbreak of war in 1939 affected the company's plans and operations as well as those of its subsidiary, West Kootenay Power. Cominco's energy requirements would again force plant extension and construction along the Kootenay River.

The production from Cominco's new chemical and fertilizer operations at Warfield — like the metals produced at Trail — acquired new significance once the country was at war. The federal government contracted to buy considerable quantities of ordnance-grade ammonium nitrate, for example, and further plant expansion was necessary to meet this order.[1] More generally, Cominco anticipated that escalating war-time demand for munitions and explosives would soon exceed its capabilities. The only solution was further growth, although Cominco was becoming anxious about the adequacy of its energy supply. In March 1940, two 25,000 horse power machines were added at Number Two plant, an extension which increased its capacity from 34,000 to 84,000 hp. A temporary power surplus resulted, but this did not last long.[2] Eighteen months later, Cominco's general manager, Selwyn Blaylock, approached the federal government with a request for help.[3]

The need for more power . . .

I believe I should call to your attention the power situation at Trail which not only concerns our metal production but also the possible production of chemicals. Increases in our possible production show that the power supply is not sufficient to take advantage of the full output which the Trail plants are capable of. At the present time, West Kootenay Power and Light Company's production is at its peak. . . . I believe it is safe to say that we are short between 30,000 and 40,000 horse power of the power required to run the whole of the Trail plants to capacity, a large part of the time. Besides, there is no spare power at any time and if a generator should be shut down for any reason, there would be a further shortage. . . .

The West Kootenay Power and Light Company has an excellent power site at Brilliant where for some years back work of a preliminary nature has been done. It would take about two years to put a plant into production at this site.

— S. G. BLAYLOCK

to C. D. HOWE, Minister of Munitions and Supply, 9 September 1941

A victim of progress: an unnamed man and two girls pose with a huge sturgeon, trapped during the construction at Brilliant in 1943.

Blaylock wrote to C. D. Howe—the "Minister of Everything" in Mackenzie King's cabinet during the war—from Montreal, site of Cominco's head office. He was soon on his way back to Trail, but not before Cominco and senior government officials had come to a draft agreement. The terms allowed for the construction of a dam at Brilliant, near where the Kootenay River flowed into the Columbia. The cost of the dam would be tax-deductible, that is, Cominco would be able to write off the cost against its net profits, which otherwise would have been subject to federal taxation. The rationale for this concession was straightforward enough: the government acknowledged that its contracts for Cominco's production of ammonia, nitric acid, and ammonium nitrate had contributed to the depletion of West Kootenay Power's surplus generating capacity. All things being equal, Cominco would not have needed the extra power for some years.

West Kootenay Power had intended to build the Brilliant plant ten years earlier but the Depression forced it to delay these plans.[4] Some preliminary work had already been done, however, and the company quickly pressed ahead with the design, once it became clear that a fifth plant was needed on the river.

> After Number Two plant extension was built—it was finished in 1940—they immediately started the designs for the Brilliant plant. Throughout this time I was in the office in South Slocan [working as] assistant to Mr. Tindale, and we had about four or five engineers and draughts-

men working [with us]. We started on the design of the Brilliant plant and it wasn't too long after that Cominco organized the construction. Mr. Chapman was still the construction superintendent. He looked after the original field work on the job. When construction got going, he was in full charge of that. . . . West Kootenay offices in South Slocan did all the design on it.

— Interview #14, Art Lambert, 4

Construction of the Brilliant plant — also known as Number Five — began in April 1942 and lasted for a little over two years. The first two turbines were in operation by August 1944, each producing 37,000 horse power. The Brilliant project differed in one important respect from the company's earlier work in the 1920s and the 1930s. Although West Kootenay Power staff designed the plant, Cominco was responsible for its construction. R. G. Anderson — West Kootenay Power's general manager following Lorne Campbell's death in 1947, and subsequently company president from 1956 to 1965 — later explained the reason for this.[5] During the course of the war, the federal government brought in a series of restrictions on both people and industry.[6] As a munitions supplier, however, Cominco was exempt from these and was able to obtain labour and material relatively easily. West Kootenay Power, on the other hand, could expect no relief from the tight constraints of war-time regulation.

The Brilliant plant commenced around that time [1942], and Mr. Stiles and I were sent back to Trail during the construction of the Brilliant plant, where I was again looking after the purchasing end. West Kootenay really had the water rights on the stream there and [had done] all the designs and plans for the Brilliant plant. During the war, West Kootenay — as a power company — had no priority for materials. Cominco was supplying metal, etc., for the war effort and they were the only people [who could] get priority [to obtain war-time material] to build a plant. That is why Cominco built the plant rather than West Kootenay.

— Interview #19, R. G. Anderson, 3

The Brilliant dam was the last one to be built along the Kootenay River between Castlegar and Nelson, and its completion brought to an end the construction era. Once the plant was fully operational, West Kootenay Power was generating electricity from ninety-seven per cent of the fall in the Kootenay River between Kootenay Lake and the Columbia.

The federal government's willingness to assist Cominco with the Brilliant dam tacitly acknowledged the company's significant contribution to the war effort. The company's role in providing heavy water —

or deuterium oxide — for atomic research during the war is additional evidence of its importance during this critical period.

In 1934, Cominco engineers had speculated on the possibility of producing heavy water, raising the matter in correspondence with the National Research Council of Canada.[7] The potential to produce heavy water was a consequence of the company's pollution control program. Cominco's ammonia plants at Warfield — built in 1929–30 as part of its strategy to recover sulfur dioxide from the smelter gases — included an electrolytic hydrogen plant; heavy water could be produced from electrolytic hydrogen.[8] West Kootenay Power played an important part in all of this — the hydrogen plant was a very heavy energy user, accounting for about a third of Cominco's consumption.[9] It was no coincidence that the only other producer of heavy water during the Second World War — in a remote part of Norway — was located at a hydro-electric plant.

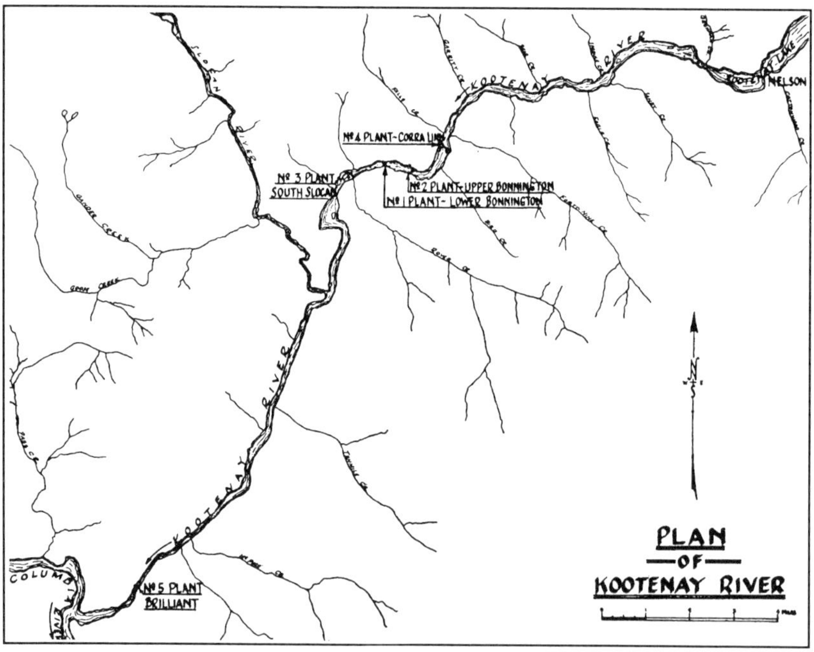

Above, a map shows the location of the five dams on the Kootenay River. The profile opposite reveals how — with the construction of Number Five plant at Brilliant — virtually all of the Kootenay River's flow was utilized to generate power. [From W. J. Tindale, "Power Development on the Kootenay River," *The Engineering Journal*, Vol. 31, #5 (May 1948), 285, 286]

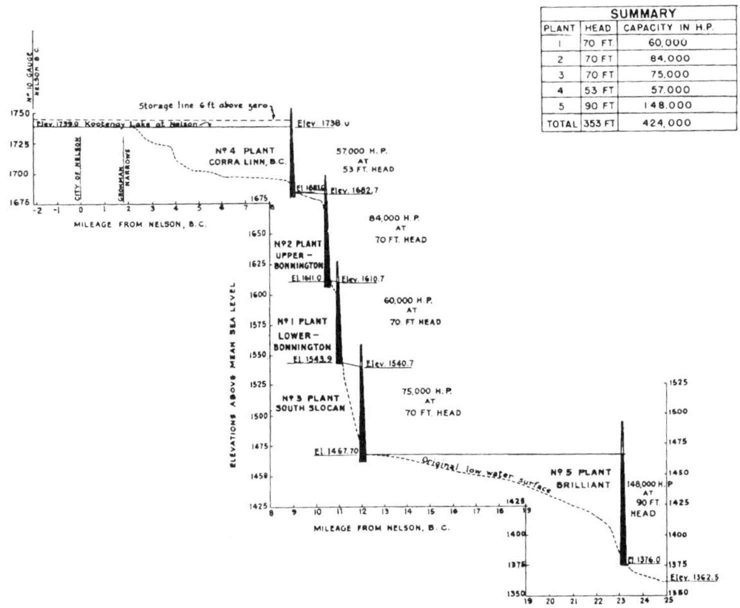

PROFILE OF KOOTENAY RIVER SHOWING HYDRO ELECTRIC DEVELOPMENTS.

In early 1941, the National Research Council of Canada wrote to Cominco's general manager requesting information about the company's ability to produce heavy water. Several months later, an American scientist wrote to ask if Cominco would be able to provide heavy water for important war-related experiments then being conducted in the United States: "we have come to the conclusion that the most rapid method of reaching a technical output of heavy water . . . at the present time is to make use of the opportunities at the Trail plant."[10] The United States was still neutral at this point—it did not enter the war until the Japanese bombing of Pearl Harbour in December 1941—and its program of atomic research was separate from the British experiments then underway. But neither Cominco nor the federal government had any qualms about providing assistance to the American atomic program. In any event, although some misunderstandings and difficulties did occur, British and American scientists tended to share the bulk of their atomic research. And increasingly, Canada was a party to these investigations. By 1943, for example, much of the British research had re-located to the laboratories of the Université de Montréal, with the assistance of the Canadian government. Ottawa also decided to acquire the El Dorado mine, the uranium ore of which was destined for the American project.[11] In Trail, Cominco's general manager signed an agreement with the American government at the end of 1942, following several months of hard bargaining. Briefly, the American government would build a heavy water plant at Trail—entailing an outlay of some two million dollars—which Cominco would operate, selling the heavy water to the Americans at cost.[12]

This war-time atomic research was a watershed in the making of the contemporary world. For better or worse, the nuclear age was dawning, and the production of heavy water had a small but significant part at this early stage. Events in Norway underlined heavy water's strategic importance during the Second World War. Like British Columbia, Norway's mountainous areas provided some excellent hydro-electric sites and Norsk Hydro operated one such plant at Vemork, west of Oslo, which produced limited quantities of heavy water. Following their invasion of Norway, the Nazis ordered Norsk Hydro to increase the plant's heavy water production, and in early 1942 demanded a second increase. The Norwegian underground passed this information on to the British government, who realised that the Nazis must be

Canada Jails Three For Sabotage Plot

Canadian industries were warned tonight to strengthen "all possible precautions" against wartime sabotage, following disclosure that police had frustrated an alleged plot to wreck the biggest smelter of its kind in the British Empire.

The plant, the Consolidated Mining and Smelting Company's at Trail, BC, is an important source of finished metal for Britain's war machine.

Provincial police arrested three "enemy aliens" who purportedly planned to "wreck" a vital unit of the huge smelter on Christmas Day. They held the suspects incommunicado and refused to reveal even their names.

The suspects were taken to Calgary for questioning by military authorities. . . .

Sergeant C. W. A. Barwis of the Nelson police detachment said the three were arrested after seizure of correspondence outlining sabotage plans. For several days there had been rumors of the plot. Police would not admit knowledge of it until the arrests were made.

Consolidated's smelter . . . had been guarded closely since the outbreak of the war.

— *New York Times*, 3 January 1940

planning to build an atomic device. To prevent this from happening, the British launched several commando raids against the Vemork plant. Not all of these were successful—and some eighty people, including thirty-four Norwegian civilians, died in the attacks—but ultimately the plant was disabled and the stocks of heavy water destroyed.[13]

West Kootenay Power was much more fortunate; no attacks disturbed its operations. "Enemy aliens" did attempt to sabotage the Trail smelter, although it's difficult to tell if this abortive plan really posed a serious threat to Cominco's operations.

Naturally enough, security precautions became the order of the day during the war. One former employee recalled that "there was a search light mounted on the roof of Number One plant. We had a job even to get into the powerhouse without a pass. . . ."[14]

Wartime along the Kootenay River

We had an organization here, it was backed by the government. . . . Pacific Coast Mountain Rangers. . . . A lot of them were World War One veterans, but most everybody working here belonged to it. Every week or so we'd have exercises, you know. They gave each one of us a rifle and we had rifle practice and sometimes the army would send in a man with sophisticated weapons and give us a try out. Machine guns and throwing these hand grenades . . . they had men guarding all these plants night and day. . . . I remember they blacked out all the windows of the power houses so there was no light showing at night. . . . The thing they were afraid of here was Japan, sending a bomber off of a ship out at sea.

— **HARRY NIXON**, a West Kootenay Power employee during the war

The only sign of hostilities in the Kootenays arrived in the unlikely form of very large paper balloons. A number of wartime employees recalled these balloons—apparently launched from Japan—which aimed to create havoc by igniting fires. Fortunately none were successful.[15]

The war's most obvious impact in the Kootenays was the sustained level of economic activity, guaranteeing full employment and steady incomes for virtually everyone in the region. Cominco was soon operating at full production, and this in turn kept West Kootenay Power's plants along the river running at full load. The only problem worrying company officials was the possibility of a power shortage, which would mean that Cominco would not receive sufficient energy to maintain its operations.

Running at Full Load

Expanded production of strategic materials at Trail increased power consumption until all generators had to run continuously. The average power generated rose from 154,000 kilowatts in 1940 to a record high of 188,000 kilowatts in 1942. The entire system was loaded to the extent that major maintenance work [at Trail] was impossible without reducing production. . . . In addition there was always the possibility that under adverse hydro conditions [i.e., low water flow] the system capability could have been reduced to an average of 145,000 kilowatts through the storage season.

— "The Cominco Story: Power," *Canadian Mining Journal*, May 1954

Overall, the war created few hardships for those who worked for West Kootenay Power. The most significant impact was a curious reversal from the earlier years of the Depression. If work had been scarce during the 1930s—and most who were employed lived in fear of losing their jobs—now the problem was a lack of mobility.

Wartime along the Kootenay River

A lot of us had our plans just abruptly changed by [West Kootenay Power's General Manager, Lorne] Campbell. He called a meeting and told us that we were valuable to the war industry, etc., and none of us should even attempt to quit our jobs and enlist. In fact, we were frozen on the job. . . . We were flatly told that if we did quit, we would be sent back to our jobs. Mr. Campbell . . . told us that we were just as important as any man in the fighting forces. . . .

— J. E. PARKER, a West Kootenay Power employee during the war

The war was clearly the period's most dramatic event and had an obvious impact on the region. Few people escaped feeling its effects, and businesses everywhere responded to its bracing stimulus. However, the war years witnessed other, more subtle developments, notably the growth of government regulation. This too influenced West Kootenay Power even if it appears far less sensational than the war itself.

From the late 1930s, the provincial government became more and more interested in overseeing various aspects of the economy. One initiative, in particular, affected West Kootenay Power: the creation of a public utilities commission just before the war and the formation of the BC Power Commission in 1945. This brought the company firmly under the control of the provincial government. West Kootenay Power's subsequent career was to demonstrate just how profound the consequences of regulation could be.

The political debates that have marked much of BC's history help to explain why and when regulation came to the province's public utilities. British Columbia is often described as a highly polarized and politicized place. Whatever the truth of this claim, some issues do recur again and again, arousing much passion, both for and against. The role and influence of big business is one such issue and has been a point of fierce debate for a century or more. During the 1898 provincial election campaign, for example, one candidate asked the crowd at a public meeting in Vancouver "whether it was anxious for its children to grow up as slaves and servants of rich monopolies or as free independent sons of the soil?"[16] The question was likely meant as a rhetorical device but public condemnation of big business and its influence on the provincial government has always been — and a century later, still remains — a familiar refrain in political debates.

Until the emergence of large forestry companies, criticisms of the power of big business typically centred on the influence that railway companies — and especially the Canadian Pacific Railway — exercised over the provincial government.[17] Since both Cominco and West Kootenay Power were CPR subsidiaries, they attracted much the same criticism. Some people, like Nelson's Mayor Houston, did not distinguish between the three companies, blaming all three for effectively preventing the emergence of a dynamic regionally-based economy.[18] Calls for government regulation or control frequently accompanied such allegations.

The demand for some form of utility regulation — such as a public utilities commission or similar regulatory device — was loudest in the Lower Mainland. This reflected popular opposition to the growing power of the BC Electric Railway Company [BCER]. The company occupied an important place in Victoria and the Lower Mainland, where it ran the street railway systems as well as providing electricity to city businesses and residences. BCER's London-based directors and their agent in British Columbia also possessed considerable political leverage, which they were not afraid to use. In 1900, for example, they arranged to have a clause added to the City of Vancouver's charter which explicitly forbid the city from competing with the BCER unless it first made a legitimate offer to purchase the company.[19] Fourteen years later, when it seemed that Premier McBride was seriously considering establishing a regulatory watchdog, one of the BCER's London principals wrote him a "Personal and Confidential" letter: "My frank opinion is that this is psychologically the wrong time to discuss the creation of a Public Utility Commission. Just at present British Columbia needs capital and the goodwill of the British investing public far more than it does a Public Utility Commission."[20] McBride took the hint and dropped the idea.

Of course, neither its high profile nor its political clout made the company any friends. "The unpopularity of the BCER," it was recently pointed out, "stemmed from a number of factors. Its lighting rates were among the highest on the continent; rapid suburban expansion in Vancouver made transit service inadequate. Above all, the company's management had a reputation with municipal leaders for driving a very hard bargain."[21] All of this contributed to the demand for some form of utility regulation, although public reports also added weight to the cause. In 1912 a Royal Commission on Municipal Government hinted at the need for regulation, while five years later Adam Shortt explicitly recommended the creation of a public utilities commission, in his published *Report of Commissioner Appointed to Investigate the Economic Conditions and Operations of the British Columbia Electric Railway Company and Subsidiary Companies*.[22] In a rather bizarre series of events, such a commission was created in the spring of 1919, only to be disbanded twelve months later.

The Liberal government of John Oliver — who assumed the premiership after Brewster's death — enacted the legislation creating the prov-

A Public Utilities Commission

Many recent circumstances have shown the need for such an authority. The Commission would perform for the province in respect to many public services the same functions that the railway commission accomplished for the Dominion in the matter of railways, Street railways, jitney services, etc. Electric and gas lighting, heating and power and telephones are among the public utilities. . . . There is no doubt the public enjoys the privilege of airing its grievances in a general sort of way, even where there is no redress. But it is desirable that the people should have a court to remedy real wrongs. . . . It is also desirable that the corporations charged with failure to perform their public duty should be able to meet these accusations before a competent tribunal.

— **PREMIER HARLAN BREWSTER**, 1917

ince's first public utilities commission, although the same government subsequently dissolved the commission when federal railway legislation moved at least some of the BCER's activities outside the scope of provincial jurisdiction.[23]

The issue of utility regulation did not re-surface until the early 1930s, when a new Liberal leader, Duff Pattullo, ousted the Conservatives and became premier. During the campaign in late 1933 he promised a public utilities commission, although the press of other business meant that the item did not receive serious attention until Pattullo's second term.[24] Perhaps the premier intended to take up the issue of regulation after his re-election in 1937 in any event, but his hand was forced by the municipalities of Victoria, Saanich, Esquimalt, and Oak Bay. They commissioned a report in 1938 that again called for the establishment of a public utilities commission.[25] Before the year was out, "An Act to provide for the Regulation of Public Utilities" was before the Legislature, passing into law in December.[26] The act gave the new Public Utilities Commission broad supervisory and regulatory powers, including the authority to revoke operating licenses. This re-constituted body was to exert far more influence than its predecessor of 1919–20.

The Regulation of Public Utilities . . .

9. Every public utility shall obey all orders of the [Public Utilities] Commission . . . and shall do all things necessary to secure observance of those orders. . . .

12. . . . No privilege, concession, or franchise hereafter granted to any public utility by any municipality or other public authority shall be valid unless approved by the Commission. . . . No public utility shall hereafter begin the construction or operation of any public utility plant or system, or of any extension thereof, without first obtaining from the Commission a certificate.

14. Where . . . the Commission determines that a public utility holding a franchise, licence, or permit has failed to exercise or make use of, or has not continued to exercise or make use of, the rights and privileges granted the said public utility by the franchise, licence, or permit, the Commission may cancel the franchise, licence, or permit . . . and . . . the public utility shall cease to operate.

— "An Act to provide for the Regulation of Public Utilities," 1938

The province's second commission shared one important feature with the first: it was largely a reaction against the BCER's dominant position in the south western corner of British Columbia.

Between one commission and the next, however, the BCER had changed. Ten years earlier, a consortium based in central Canada purchased the company as well as its subsidiaries, turning them over to a holding company, the BC Power Corporation.[27] Some things, however, remained unchanged: complaints against the BCER had continued and explain why one of the first acts of the Public Utilities Commission was to examine the company's rates and operations, an exhaustive study that was not completed until 1943. The commission sided with electrical consumers, concluding that rates were too high. As a result of its recommendations, some BCER customers enjoyed periods of free power, from one to two months, for several years. Nor did West Kootenay Power escape the eye of the commission. After its study of the BCER, the commission's attention turned to the province's other utilities. West Kootenay Power saw the writing on the wall, however, and without waiting for an order from the commission, the company voluntarily reduced its rates by 20%.[28] It would be a mistake, however, to see the activities of the Public Utilities Commission as simply a means to control the BCER and West Kootenay Power. Its creation, and the sweeping powers that it was given, reflected a growing recognition that electricity now played a vital role in the province. As far as the government was concerned, it had become too important to be left in the hands of private companies.

The government appreciated electricity's significance and knew that many people living outside the Lower Mainland did not have ready access to power. The Rural Electrification Committee — created in 1943 — looked into the problems of electrical supply to these less populous regions. In 1944 the committee concluded that the cost of power in rural areas was high, and that both the quality of service and the level of use was relatively low.[29] In January of the following year it produced a second report, recommending ways to improve and extend electrical supply to rural parts of the province.[30] The government responded with legislation, "An Act to provide for improving the Availability and Supply of Electrical Power," which created a new government-owned utility, the British Columbia Power Commission.[31] The province had entered the electrical supply business.

Although another sixteen years passed before the government startled everyone by announcing plans to take over the BCER and create BC Hydro, it had taken the first steps down that road.

The owners of West Kootenay Power followed these developments nervously. Cominco was not primarily an electrical supplier, nor did it have any interest in seeing West Kootenay Power become a public utility. The role of West Kootenay Power, even prior to its purchase by

> ### *Powers of the Commission*
>
> To generate and supply power. . . .
>
> To develop power-sites, power projects, and power plants. . . .
>
> To build, make, or establish every kind of structure, excavation, or installation suitable for the development or operation of power-sites, power projects, or power plants. . . .
>
> To acquire by purchase, lease, or otherwise the whole or any part of the real or personal property of any person who owns, operates, or controls a power-site, power project, or power plant. . . .
>
> On the recommendation of the [British Columbia Power] Commission the Lieutenant-Governor in Council may . . . authorize the Commission . . . without the consent of the owner. . . . To take by expropriation . . . any property, real or personal, and any power-site, power project, or power plant.
>
> – "An Act to provide for improving the Availability and Supply of Electrical Power," 1945

the CPR in 1905, was that of an energy supplier for the mining industry. Cominco's chief aim after acquiring the company in 1916, as the company president explained to shareholders several years later, was to guarantee "an adequate supply of electrical energy . . . at a reasonable cost."[32] West Kootenay Power had certainly done this for Cominco, although up to 1946 the utility had not found many other customers for its electricity. From 1927 to 1946, for example, outside sales had grown from 4,800 kw to 5,500 kw. Averaged out over the twenty year period, this represents a very modest annual increase of perhaps a percentage point or two a year. In light of West Kootenay Power's minor role as a electrical supplier beyond the industrial complex at Trail, and with the government becoming more interested in public utilities, Cominco concluded that the time had come to re-structure. It would take over most of the Kootenay River power plants and merge them with its own operations, while West Kootenay Power would become the public power company, supplying electricity to the other businesses and homes of the region. West Kootenay Power would be left with the Number One plant, whose output seemed capable of meeting the region's power requirements for the forseeable future. With the approval of the Public Utilities Commission, West Kootenay Power sold Number Two, South Slocan, and Corra Linn to Cominco. Even with its power generation reduced to the Number one plant, the company still had an energy surplus and was able to sell power back to Cominco.

> In 1947 Cominco bought all the plants, except the Lower Bonnington Plant, from West Kootenay. Of course, they already owned the Brilliant Plant. They bought the other plants with the consent of the Government and the Public Utilities Commission at that time, because the [Number One] plant appeared to have lots of capacity to serve the public for a good many years . . . the load growth at that time was probably one or two per cent.
>
> — Interview #19, R. G. Anderson, 6

Domestic power consumption subsequently took off and forced Cominco to modify these plans. Nonetheless, the 1947 division signalled an important change of focus for Cominco. It also demonstrates how utility regulation encouraged the company to re-think its relationship with West Kootenay Power and anticipates the re-structuring of the 1980s when Cominco took the final step, completely severing its connection with West Kootenay Power.

Another significant event occurred in 1947, one that reflected West Kootenay Power's past rather than foreshadowing its future. In the

Lorne Campbell, 1871–1947

He was the perfect bull of the woods in looks and in voice, and he had this big booming voice and this purple red face (especially if he got mad), and he'd come storming in and everybody would jump to attention. This is literally. . . . As far as his own family was concerned, he finally got so miserable they couldn't live with him, so Mrs. Campbell and family, they moved to Victoria and Lorne stayed here. . . . He didn't have any friends toward the end. My father used to tell me. . . . "I think I'm his only real friend," and [that was] only because they had been together so long. He understood the man, and when I got so annoyed and [would] throw up my hands, he would remind me to not take it too seriously, "Mr. Campbell really likes you, that's the way he is."

. . . They would almost draw straws to see who would phone him down there [at Cominco], if there was a contentious point. One of the things that bothered people a little bit when Cominco took over [the three plants in 1947, after Lorne Campbell's death] was the rush. There was a big neon sign on Number Three plant [at South Slocan] that says "Consolidated Mining and Smelting Company." That neon sign used to say "West Kootenay Power and Light Company" and within a week it was changed, and I think a lot of people thought it was an affront to Lorne Campbell. You could dislike him or anything you like, but he had contributed a lot to the area [and] to the company, and they felt this was done in too much of a hurry. However, it was done."

— JACK MCDONALD

spring of the year, the company's general manager, Lorne Campbell, died. For fifty years, he had guided West Kootenay Power's operations and his death signalled the end of an era. By all accounts, Campbell was a tough customer who rarely saw any difference between his personal life and that of the company, and had almost as much trouble making the distinction in other people's lives. When twenty-six retired employees of West Kootenay Power were interviewed in the mid-1970s, each had stories to tell of Lorne Campbell. By and large they were not particularly fond memories of the man, but all reflected a sincere respect for him. One of the people closest to him stated bluntly that "To me personally, when Lorne Campbell died, the history of West Kootenay came to an end. It was no longer the same company [after his death], it was just a wing of Cominco."[33]

Lorne Campbell's death, coinciding as it did with the sale of the three plants to Cominco, meant that the character of West Kootenay Power would be very different in the post-war years.

NOTES

1 "The Cominco Story: Chemicals and Fertilizers Division," *Canadian Mining Journal*, Vol. 75, No. 5 (May 1954), 287.

2 W. J. Tindale, "Power Development on the Kootenay River," *Engineering Journal*, Vol. 31, #5 (May 1948), 288. He points out that the Number Two Extension, as it was called, "raised the capacity on the river to 276,000 hp. and created a temporary balance of power above the demand."

3 S. G. Blaylock to C. D. Howe, 9 September 1941, in Canada, House of Commons, unpublished *Sessional Papers*, 3rd session of the 20th Parliament, 30 January-17 July 1947, No. 244, RG14, D2, Vol. 519, National Archives of Canada. Detail in the following paragraph is largely from this same file (No. 244), notably the formal agreement drawn up by Cominco and the federal government, dated 20 April 1942. As the accompanying correspondence in the file suggests, the press of war-time and the convoluted nature of the agreement had delayed the signing date; to all intents and purposes, the federal government had agreed to the terms in September 1941.

4 See for example "News of the Field: Plan Fifth Project on Kootenay River," *Power*, Vol. 73 (2 June 1931): 898, and International Joint Commission, *The Kootenay Valley: A report on certain cases involving reclamation and the development of water power in the valley of the Kootenay River, under the terms of article IV of the Treaty of January 11, 1909*, Ottawa: King's Printer, 1936, 105.

5 Interview #19, R. G. Anderson, 3. Grant Tindale offered much the same analysis: see Interview #18, 3-4.

6 The best account of the government's activities during the war remains J. L. Granatstein, *Canada's War: The Politics of the Mackenzie King Government, 1939–1945*, Toronto: Oxford University Press, 1975.

[7] This account draws on information provided by C. D. Andrews in his article, "Cominco and the Manhattan Project," *BC Studies*, No. 11 (Fall 1971): 51-62. However, Andrews focuses exclusively on events at Trail and as a consequence overlooks much important information. For the bigger picture, see Brian L. Villa, "Alliance Politics and Atomic Collaboration, 1941–1943," in Sidney Aster, ed., *The Second World War as a National Experience*, Ottawa: Canadian Committee for the History of the Second World War, 1981, 148-62, and the first chapter of Robert Bothwell's *Nucleus: The History of Atomic Energy of Canada Limited*, Toronto: University of Toronto Press, 1988. Three other important sources are C. P. Stacey, *Arms, Men and Governments: The War Policies of Canada, 1939–1945*, Ottawa: Queen's Printer, 1970; Margaret Gowing, *Britain and Atomic Energy, 1939–1945*, London: Macmillan, 1964; and Richard G. Hewlett & Oscar E. Anderson, Jr., *The New World, 1939/1946: A History of the United States Atomic Energy Commission, Volume I*, University Park: Pennsylvania State University Press, 1962.

[8] For descriptions of this process, and more general comments on the hydrogen plant, see "The Cominco Story: Chemicals and Fertilizers Division," 292-94; *The Story of the Consolidated Mining and Smelting Company of Canada, Limited*, Trail: Trail Times Ltd., 1938?, 15-7; and A. G. Dickinson, "Utilization of Electric Power at the Metal and Fertilizer Plants, Trail, BC," *Engineering Journal*, Vol. 20, No. 9 (September 1937), 722-23.

[9] By the late 1930s, the hydrogen plant accounted for some 80% of the power used at Warfield, which translates into about a third of Cominco's total consumption. See Dickinson, "Utilization of Electric Power at the Metal and Fertilizer Plants," 719, 723.

[10] Dr. Hugh Taylor to S. G. Blaylock, 21 June 1941, quoted in Andrews, "Cominco and the Manhattan Project," 53.

[11] See Bothwell, *Nucleus*, esp. 3-53; Bothwell, *Eldorado: Canada's National Uranium Company*, Toronto: University of Toronto Press, 1984; and Villa, "Alliance Politics and Atomic Collaboration, 1941–1943."

[12] See Andrews, "Cominco and the Manhattan Project," esp. 59-61. As Villa suggests, the charge that these arrangements were done without the knowledge of the Canadian government is untenable. Prior to negotiating with the Americans, Blaylock had written to C. D. Howe, advising him of Cominco's plans. Howe assured him that "I am familiar with the whole project to which you refer," and suggested that Blaylock "go on with whatever representatives of the United States Government may wish you to do in this connection" (Howe to Blaylock, 31 July 1942, quoted in Andrews, 59; see also Villa, "Alliance Politics and Atomic Collaboration, 1941–1943," 146-47).

[13] See Martin Gilbert, *Second World War*, London: Weidenfeld and Nicolson, 1989, esp. 319, 379, 404, 475, 500, as well as the autobiographical account by Knut Haukelid, *Skis Against the Atom*, London: William Kimber, 1954, translated by F. H. Lyon, with an Introduction by Major-General Sir Colin Gubbins.

[14] Interview #4, Jock Hawkins, 5. See also Interview #5, Harry Nixon, 3-4; Interview #23, J. E. Parker, 9; and Interview #24, Jess Ridge, 8.

[15] See the comments in Interview #5, Harry Nixon, 4-5, and in Interview #4, with George Helbecque, Ivor Jones, Jock Hawkins and Davie McDonald, 6.

[16] *Daily News-Advertiser* (Vancouver), 29 June 1898, 2. The power of big business was one of the central issues in the 1898 provincial election, and I discuss this briefly in *Roaring Days*, 44-46.

[17] In 1901, for example, E. V. Bodwell—the prominent Victoria lawyer who subsequently acted for the Cascade Water, Power and Light Company in its 1906 suit against

West Kootenay Power, described in Chapter Three — denounced the CPR's influence on the provincial government, provoking an angry response from the Premier on the floor of the Legislature. (See the account of Bodwell's speech in "Endorsed The Road," *Victoria Daily Times*, 20 March 1901, 6, and for Premier Dunsmuir's response to Bodwell, see "Provincial Parliament," *Victoria Daily Times*, 21 March 1901, 3.) The following year, Smith Curtis, Rossland's MLA and a former cabinet minister, charged Premier Dunsmuir as well as various cabinet ministers with giving preferential treatment to railway companies, and demanded a public inquiry: see "Charges And Defence," *Victoria Daily Times*, 20 March 1902, 4 (editorial); "Proceedings of the Legislature," *Victoria Daily Times*, 20 March 1902, 3, 6. For the subsequent public inquiry, see "The Commission On Railway Deal," *Victoria Daily Times*, 27 March 1902, 1, 8; "Royal Commission Hears Evidence," *Victoria Daily Times*, 29 March 1902, 3, 8.

18 For examples of these charges, see the discussion in Chapter Four.

19 See *Statutes of the Province of British Columbia . . .*, 1900, Chapter 54, "An Act to Revise and Consolidate the Vancouver Incorporation Act." The relevant section reads "the [City of Vancouver] Council shall not . . . pass any by-law for the purpose of purchasing, acquiring, constructing, operating or maintaining any works similar to those now carried on by the British Columbia Electric Railway Company, Limited . . . until the Council has, by by-law, fixed the price which they will offer for the property of the company" (324). For the ways in which the company achieved this remarkable clause, see Christopher Armstrong and H. V. Nelles, *Monopoly's Moment: The Organization and Regulation of Canadian Utilities, 1830–1930*, Philadelphia: Temple University Press, 1986, 96-100. On the company's history generally — in addition to the relevant sections of Armstrong and Nelles's book — see the various studies by Patricia Roy: "The British Columbia Electric Railway Company, 1897–1928; A British Company in British Columbia," PhD thesis (History), University of British Columbia, 1970; "Regulating the British Columbia Electric Railway: The First Public Utilities Commission in British Columbia," *BC Studies*, #11 (Fall 1971): 3-20; "The Fine Arts of Lobbying and Persuading: The Case of the British Columbia Electric Railway, 1897–1917," in David S. Macmillan, ed., *Canadian Business History: Selected Studies 1497–1971*, Toronto: McClelland and Stewart, 1972, 239-54; "The British Columbia Electric Railway and Its Street Railway Employees: Paternalism in Labour Relations," *BC Studies*, #16 (Winter 1972–73): 3-24; and "Direct Management from Abroad: The Formative Years of the British Columbia Electric Railway," *Business History Review*, Vol. 47, No. 2 (Summer 1973): 239-59.

20 R. H. Sperling to McBride, 8 January 1914, quoted in Armstrong and Nelles, *Monopoly's Moment*, 207.

21 Armstrong and Nelles, *Monopoly's Moment*, 207.

22 See the *Report of Commissioner Appointed to Investigate the Economic Conditions and Operations of the British Columbia Electric Railway Company and Subsidiary Companies*, Victoria: King's Printer, 1918, 46-49. The earlier Report of the Royal Commission on Municipal Government was published in the *Sessional Papers of British Columbia for 1913*, L1-L18. Note also the discussions of both reports in Armstrong and Nelles, *Monopoly's Moment*, 207, 258-60, and Roy, "Regulating the British Columbia Electric Railway," 4, 7-8.

23 For illuminating accounts of BC's first Public Utilities Commission, see Armstrong and Nelles, *Monopoly's Moment*, 206-10, 254-62, and Roy, "Regulating the British Columbia Electric Railway."

24 See the comments in Mary Doreen Taylor, "Development of the Electricity Industry in British Columbia," MA thesis (Geography), University of British Columbia, 1965, 84, and Robin Fisher, *Duff Pattullo of British Columbia*, Toronto: University of Toronto Press, 1991, 295-96. Fisher's book includes a thoughtful discussion of Pattullo's ideas and policies going into the 1933 election (212-16), although Margaret Ormsby's earlier article remains useful: see "T. Dufferin Pattullo and the Little New Deal," *Canadian Historical Review*, Vol. 43, No. 4 (December 1962), esp. 282-85. For a brief and insightful overview of the province's political scene, see Robin Fisher and David J. Mitchell, "Patterns of Provincial Politics Since 1916," in Hugh J. M. Johnston, ed., *The Pacific Province: A History of British Columbia*, Vancouver: Douglas & McIntyre, 1996, 254-72.

25 See Taylor, "Development of the Electricity Industry in British Columbia," 84-85.

26 See *Statutes of the Province of British Columbia . . .*, 1938, Chapter 47, 185-221; also Taylor, "Development of the Electricity Industry in British Columbia," 85-86. As Armstrong and Nelles suggest in *Monopoly's Moment*, effective utility regulation came very late to British Columbia, particularly if one compares the province to other populous regions of Canada. In 1973, the BC Energy Act replaced the Public Utilities Commission with the BC Energy Commission.

27 On the sale of the BCER, see Roy, "The British Columbia Electric Railway Company, 1897–1928," 294-351; H. William Tieleman, "The Political Economy of Nationalization: Social Credit and the Takeover of the British Columbia Electric Company," MA thesis (Political Science), University of British Columbia, 1984, 27-31; and Armstrong and Nelles, *Monopoly's Moment*, 300.

28 Taylor, "Development of the Electricity Industry in British Columbia," 87.

29 British Columbia, *Progress Report of the Rural Electrification Committee As of January 24th, 1944*, Victoria: King's Printer, 1944, 9, 81-82. The committee pointed out that "The use of electricity has nowhere reached saturation and in British Columbia it has a long way to go" (50).

30 See British Columbia, *Report of the Rural Electrification Committee As of January, 1945*, Victoria: King's Printer, 1945, esp. 29-31, 44-5.

31 For the act, see *Statutes of the Province of British Columbia . . .*, 1945, Chapter 27, 85-114.

32 Quoted in British Columbia, *Annual Report of the Minister of Mines for 1918*, K177.

33 Interview #2, Jack McDonald, 11; note also the comments in Interview #7, R. W. Diamond, 1-2.

CHAPTER SEVEN

The Post-War World: *Affluence, Growth, and the Politics of Power*

Almost as soon as Cominco down-sized West Kootenay Power in 1947, domestic energy consumption began to soar. Two things contributed to this growth: a population boom in the region, particularly in the Okanagan Valley, as well as the general affluence of the period, which encouraged a dramatic release of pent-up consumer demand. With the Depression and war-time restrictions finally behind them, people surrounded themselves with an array of electrical appliances. Cominco had not foreseen this growth in domestic electrical consumption when it had cut West Kootenay Power's generating capacity to the Number One plant at Lower Bonnington. The arrangement soon proved unworkable.

The growing size and affluence of the region's population signalled a series of other changes. Automobiles became common, people were more mobile, and employment opportunities beckoned. West Kootenay Power's earlier approach to labour relations—marked by Lorne Campbell's gruff and paternalistic style—became out-dated, while changes in the province's labour laws meant that the company could no longer recognize its own inhouse "union." Soon it had to deal with a local of the American-based union, the International Brotherhood of Electrical Workers. Beyond the region, the election and long rule of W. A. C. Bennett and his Social Credit party were to influence the company, notably with the creation of BC Hydro in 1962. The provincial government's decision to go into the energy business meant that it became West Kootenay Power's competitor. This competition was far from equal, for BC Hydro enjoyed a number of advantages that West Kootenay Power could not hope to match.

In the years following the Second World War, Canadians came to regard electrical appliances as necessities. In 1947 the only really common electrical gadget found in a typical home was the radio. (Ninety per cent of Canadian homes — the figure used by Statistics Canada to determine the saturation point — had radios by 1947.) Refrigerators were next: although less than a third of Canadian homes boasted a fridge after the war, by 1960 they too had become a standard fixture. Television followed in 1963, and then the telephone in 1966.[1] The proliferation of electrical appliances, coupled with a population boom, meant that domestic power consumption rose dramatically in the 1950s and 60s.

The sense that electricity had become a necessary part of everyday life encouraged the provincial government to consider ways in which all British Columbians could enjoy its benefits. The government's approach took a new form with the election of a Social Credit government in 1952. The new government marked a dramatic shift in the political life of the province, ending the dominance of the Liberal and Conservative parties — the political reflections of BC's establishment — which had governed as a coalition since 1941. Prior to 1941, either one or the other held the reigns of power. W. A. C. Bennett and his Social Credit party, however, drew their support largely from outside the traditional base of the two established parties.[2] Social Credit remained in office for twenty years, forcing the province's urban elite to come to terms with Bennett, an expansive business person from the Okanagan. The new premier's particular genius was to reconcile his party's sometimes volatile rank and file with the more sedate aims of the urban business class. His skill as a political strategist, combined with his benign populism, meant that he was always able to attract a substantial number of voters from the province's middle and working class.

Bennett's election in 1952 coincided with a surge of economic activity in the province. This boom produced welcome revenue for government coffers, but it also gave a particular stamp to the Bennett era. His was to be a rule of development, of prosperity. For the next two decades, he worked assiduously to ensure that the economic boom continued unchecked. British Columbia's prosperity in the post-war era was hardly unique. Most economies in the developed world enjoyed a period of affluence, although it is hard to imagine that many other jurisdictions had a government as dedicated to development. Bennett's commitment to expansion, particularly outside the Lower Mainland, was such that he was always enthusiastic about power projects, and the bigger the better. These were heady times, and the general mood of confidence and optimism meant that proposals were not slow in coming. Several involved projects in the Kootenays, although to make sense of these we need first to touch briefly on a series of discussions concerning the Columbia River basin, talks that dated back to the closing years of the Second World War.[3]

In 1943 the American government became interested in further development of the Columbia River basin. War-time energy demand was escalating rapidly, suggesting that even the prodigious output of the Grand Coulee and Bonneville dams might one day prove inadequate. The most efficient approach to long-range planning for the river would necessarily consider its entire length. Like its tributary, the Kootenay, the Columbia was an international river, with headwaters in British Columbia. Just under 40% of the length of the river itself lies north of the border, where about half of its total fall occurs. Although British Columbia accounts for 35% of the total flood water, much of the Columbia basin lies in the United States (only 15% is in Canada).[4] Given the river's course across the border, the International Joint Commission would be involved in any plans for its development, and in 1943 the American government approached Canada with a request that the issue be referred to the IJC. Ottawa had no objections and so the whole question of the Columbia's future development went to the International Joint Commission.

Members of the IJC had little direct expertise in engineering or the other fields that would be involved in devising a realistic plan for the future of North America's fourth largest river. Consequently they assembled an advisory committee to study the matter. This was not a speedy process: fifteen years were to pass before the International Columbia River Engineering Board produced its report. Before the report was finished, the issue acquired a degree of urgency. In the late spring of 1948, the Columbia River overflowed its banks in spectacular — and disastrous — fashion. Sandbags were needed to stop water from flooding West Kootenay Power's Number One plant, while parts of Trail were under water.[5] This paled in comparison with events downstream, however: more than forty people lost their lives in the United States and property damage topped a hundred million dollars.[6] Such devastation naturally led to discussions about ways to control the flow of the river and prevent any repetition of such floods.

As a consequence of the 1948 flood, the American government brought forward a proposal for an upstream dam on the Kootenay River, in Libby Montana, which would help to avert flooding in the future. This went to the IJC in early 1951, but the dam — which would influence parts of both Montana and British Columbia — raised some difficult issues. No one questioned the potential benefits of the Libby

Dam, but the very prospect of these benefits raised the thorny question of who was to get what. For example, if flooding land in British Columbia gave certain advantages to American residents living downstream, in terms of flood protection and/or energy supply, what formula could be used to determine these downstream benefits? What could be considered reasonable compensation? This issue of downstream benefits was to recur again and again in the lengthy negotiations surrounding the development of the Columbia River. In the short term, however, difficulties in resolving the issue led to the shelving of the Libby proposal in 1953.[7]

Shortly after the International Joint Commission received the American proposal for the Libby Dam, Cominco submitted another. The company wanted to build a dam on the Pend d'Oreille River, close to its junction with the Columbia and downstream from Trail, adjacent to the American border. The proposed Waneta dam and reservoir did not encounter the same opposition that greeted the Libby proposal and in 1952 the IJC gave its approval.[8] The plan for a dam at Waneta was not new: West Kootenay Power had anticipated building it in the late 1920s, before the Depression brought a halt to the company's dreams of expansion and growth. (In fact, the fourth floor of the company's office building in Trail had been designed as a space for engineering staff to work on the project.[9]) Twenty years later, things were different. West Kootenay Power's re-structuring in 1947 narrowed its horizons and the company did not get the chance to build Waneta. Instead, Cominco awarded the construction contract to the American firm of Stone and Webster. Ownership of the Waneta dam would remain with Cominco, although West Kootenay Power assumed the management of the plant as well as responsibility for the transmission lines that carried its power to the larger system.

The Waneta plant began producing power in late 1953, with the installation of its first two generators. (Two others were later added, in June 1963 and September 1966.[10]) The addition of Waneta's considerable generating capacity gave Cominco an energy surplus. The company worked out a tentative deal to sell this extra power to a Washington utility, but the federal government's objections prevented the sale from going ahead.[11] Ottawa's reluctance to approve power sales to the United States subsequently became a controversial feature in the discussions surrounding the Columbia River Treaty.

Meanwhile, back in Victoria, two American proposals for dams on the Columbia landed on Premier Bennett's desk in 1954. As with the Libby proposal, both would result in flooding certain parts of southeastern BC, but they also promised some potential benefit for the gov-

The escalating demand for power in the post-war years resulted in the construction of numerous new transmission lines across the province.

ernment. Premier Bennett was particularly interested in the plans of the American company, the Kaiser Aluminum and Chemical Corporation. The Kaiser interests proposed constructing a dam on the Arrow Lakes and giving the province a percentage of the downstream benefits: 20% of the power generated by the plant. To the provincial government, this seemed a relatively easy way to earn a tidy annual sum. Not only that, but the plan called for no expenditure on the government's part. In the autumn of 1954, Premier Bennett and his Minister of Lands and Forests — Bob Sommers — happily announced that his government had signed a tentative agreement with Kaiser.[12]

The provincial government soon discovered that others did not share its sunny assessment of the Kaiser plan. An avalanche of criticism descended upon Victoria, from the press, the opposition, and the federal government. Although Bennett was not a person easily dissuaded from pursuing his chosen path, Ottawa's opposition was hard to ignore. A major battle between the two levels of government was not long in coming.

Political history in Canada is often little more than the history of federal-provincial relations writ large. The fate of the Kaiser plan—indeed, the whole twenty-year history of the Columbia River treaty—illustrates this point only too well. From Ottawa's perspective, the Kaiser plan threatened to short-circuit the lengthy study of the Columbia River's future begun by the International Joint Commission a decade earlier; Ottawa also suspected that the deal was too favourable to the American company. The Liberal government of Prime Minister St. Laurent swiftly introduced a bill—the International Rivers Improvement Act—that effectively killed the Kaiser proposal. Passed in the summer of 1955, the act stipulated that no one could construct a dam on an international river without a federal license, that is to say, without the approval of Ottawa.

The government in Victoria did not take kindly to Ottawa's action. As far as it was concerned, the rivers of British Columbia, and any plans for their development, were a provincial affair; it saw the federal government's new law as an infringment on provincial jurisdiction. Premier Bennett and his colleagues also took exception to the none too subtle hints that they were an inexperienced bunch prepared to sell out the province's interests at the blink of an eye. For the moment, the provincial government knew that it had been beaten—unfairly beaten, too, it would claim—but it was prepared to wait for other opportunities. These were not long in coming.

Although the next project to make headlines in British Columbia did not concern the Kootenays or its rivers, it was to exert a profound influence on events there. It all began in the spring of 1956, at a reception in the province's London House, when the representative of a Swedish industrialist learned of the tremendous potential of northeastern British Columbia.[13] Before too long, Axel Wenner-Gren sent two associates to the province, where they met with the Premier. The outcome of these meetings was a proposal from Wenner-Gren to embark on a massive development project, formally detailed in a letter of agreement between the government and Wenner-Gren, signed in the autumn of 1956.[14]

When Wenner-Gren's ambitious scheme became public the following spring, it attracted a good deal of attention, little of which was positive. The opposition and a number of the urban papers denounced the plan alternatively as a massive give-away of the province's resources and/or as an un-attainable pipe dream with little chance of success. Whatever the motives of the provincial government—notably those of Premier Bennett, firmly in charge of the government's energy policy — the Wenner-Gren proposal focused attention on the enormous energy potential of the Peace River. As the discussions of the Columbia River's future became more and more drawn-out and convoluted, an impatient Bennett considered what might be achieved in north-eastern British Columbia. Bennett was not a person who welcomed delays or dissent, and he knew development in the Peace River district would not be hampered by jurisdictional squabbling.[15]

Wenner-Gren formed the Peace River Power Development Company to inaugurate a large hydro-electric development in northeastern BC.[16] Before long, the company's plans encountered a major stumbling block. No one could question the vast hydro-electric potential of the northern river system, but essential to the success of the company's project was a guaranteed market for the considerable energy that it was going to generate. To raise the necessary capital, long-term energy contracts had to be in place, tangible evidence for would-be investors of the plan's financial feasibility. BC Electric, with its parent company, the BC Power Corporation, was the province's leading energy company, and the obvious customer for Peace River power. Although quite interested in the Peace River project — it was represented on the board of the Peace River Power Development Company — BC Electric had already made plans for its future power supply. When pushed by Bennett, the company flatly refused to sign any long-term contract to purchase Peace River power. But the premier was not easily dissuaded from his plans for northern development. Increasingly he had come to see the Peace River power project as a keystone of his government's development strategy, something that dovetailed perfectly with his vision of the province's future as well as providing a lever with which to apply pressure on the federal and US governments, in terms of the Columbia River talks.

Bennett's next move caught everyone by surprise. In one of his most controversial and unexpected acts, the premier announced in the provincial legislature on 1 August 1961 that he intended to take over both BC Electric and the Peace River Power Development Company. With hindsight, the move makes a good deal of sense, but at the time it provoked denunciation and disbelief. Since the nationalisation of BC Electric was a long-standing goal of the opposition, the plan passed without a dissenting vote in the legislature. On other fronts, the response was harshly critical. The financial press hurled insults and threats at Victoria, but Bennett appeared unperturbed. This wasn't about ideology, he explained, it was about development, about the province's future. He went on to point out that for years people had accused his government of being subservient to BC Electric. He had always denied this, and now here was proof of his independence. As premier, he would let no one — not Ottawa, not big business, not unions, not the opposition — stand in the way of his government's plans, which would benefit all British Columbians. Ever the affable populist, he smiled blandly amidst the fury and let the storm break over him.[17]

When Premier Bennett introduced the bill that effectively nationalized BC Electric in August 1961, he was heard to remark that West Kootenay Power would be next.[18] Bennett and his constituents in Kelowna were supplied by West Kootenay Power, of course, and this in itself may help to explain why the company avoided the fate of BC Electric and the Peace River Power Development Company. People in the Okanagan were well aware of the fact that they enjoyed far cheaper power than consumers in the Lower Mainland. If nationalization brought with it higher rates, Bennett's chances of re-election might have been threatened.[19] But another reason is probably more important in explaining West Kootenay Power's survival as an independent utility, outside of the new BC Hydro system.[20] Bennett had no wish to end private power generation in the province and that was not his motive in nationalizing BC Electric. Rather, the move reflected a larger political objective: he wanted to see both the Columbia and the Peace developments go ahead, what he liked to refer to as his "two river policy." This was, unquestionably, a grandiose scheme and one near and dear to Bennett's heart. "It is the largest and brightest concept anywhere today," he boasted, ". . . too big for little persons to grasp."[21] While many obstacles stood in the way of Bennett's two river policy, West Kootenay Power was hardly one of them; success did not require its nationalisation. Personal memoirs — such as those of Robert Bonner — tend to support this analysis.

An evolution of analysis

We felt we had no alternative but to nationalize the BC Electric to make it a vehicle for electrical energy policy . . . there was a growing realization that the Two River Policy was probably beyond the capacity of private interests to deal with, and when we started to become more heavily involved with the Columbia, there was no way that we could control the pressure on negotiations without having the Peace development under our control as well. It was an evolution of analysis which gradually brought us into the picture: we had to have both the Peace and the Columbia. . . .

— **ATTORNEY GENERAL ROBERT BONNER**[22]

The reason Bennett raised the issue of West Kootenay Power at all, when discussing nationalization, likely reflected his ostensible justification for the controversial move. This involved a feud with Ottawa over taxation: in effect, an attempt to win public support through reviving an old provincial theme of "fight Ottawa." Bennett had argued on other occasions — although no one took him very seriously — that if the federal government did not return to the province the tax revenues that Ottawa collected from private utilities, he would have no option but to nationalize the utilities, out of fairness to consumers. Publicly-owned utilities paid no income tax and Bennett claimed that this gave them an unfair advantage. After all, consumers who relied on electricity from a private utility were obliged — through their power bills — to assist the company pay its taxes. However, the focus on taxation was largely window-dressing; as David Mitchell noted in his biography of Bennett, "this was clearly more a rationalization than an explanation."[23]

While the controversy raged over the government take-over of BC Electric, Bennett continued to participate in the Columbia River negotiations. Diefenbaker and his federal Conservatives had somewhat precipitously signed the Columbia River treaty in January 1961. Anxious to make what political capital he could out of the Columbia River talks, Diefenbaker flew to Washington in the dying days of the Eisenhower administration to secure the presidential signature on the treaty from the outgoing Republican incumbent, before President Kennedy formally took office. A few days earlier, however, Bennett had conveyed his misgivings about the treaty to Ottawa.[24] A shrewd political strategist, Bennett knew that he had won an advantage in the ongoing debate over the Columbia, which pitted the federal government against its provincial counterpart. The signatures of the Canadian prime minister and the US president on the treaty would amount to very little if the province was unwilling to accept its conditions. The federal government could negotiate any treaty it liked, but no dams could be built on BC's rivers without the approval of Victoria. "Seldom in Canada's history," noted one Vancouver commentator, "has a Canadian Prime Minister so trustingly placed his neck in a headlock by a provincial premier as Mr. Diefenbaker appears to have done to Premier Bennett's obvious strategic advantage."[25]

Bennett had no objections to a Columbia River treaty. Indeed, the treaty was essential to his two river policy. His concern was with the treaty's bearing on his overall energy strategy: he wanted to ensure that it did not hinder the development of the Peace. Two things, in particular, caused him anxiety. The first was the federal government's policy regarding electrical sales to the United States: its opposition to long-term contracts involving the export of electricity went back many years.[26] This policy posed a potentially fatal threat to the Peace River project. If Columbia River power could not be sold in the US, its only other market would be the Lower Mainland. However, Bennett knew that the Vancouver region could not absorb power from both the Columbia and the Peace. It followed that failure to sell the Columbia power to the US would render the Peace River project redundant, and that was something that Bennett was not prepared to accept. The other issue that concerned him was the treaty's financial arrangements: how much money was to flow into his government's purse. As far as Bennett was concerned, any money earned by BC's rivers belonged to the province. Not only did he expect the money from the sale of the Columbia's downstream benefits, he also had very definite ideas about the price. The instability of the federal government throughout the early 1960s helped Bennett to achieve most of what he was after.

The June 1962 federal election ended Diefenbaker's majority in the House of Commons. His minority government after the election relied on the support of the federal Social Credit to remain in office. Four months later, Ottawa formally abandoned a long-standing policy, with the announcement in the Throne Speech that "large-scale, long-term contracts for the export of power surpluses . . . should now be encouraged."[27] One knowledgeable journalist reported at the time that the rumour in Ottawa was that federal Social Credit MPs were keeping the Tories in power so that they could ratify the Columbia treaty.[28] However, the Conservative government did not survive long enough to carry this out. A second federal election, in the spring of 1963, ousted Diefenbaker and his Conservatives. Lester B. Pearson and the Liberals formed a minority government, one that was to prove a good deal more durable than Diefenbaker's.

Several months later, on 30 September 1963, Bennett and Social Credit were re-elected in British Columbia, after a hard-fought campaign that centred on "that complex and difficult subject hydro-electric power."[29] Feeling vindicated by the results of the election, Bennett confidently began serious discussions with the new federal government over possible changes to the Columbia River treaty. Unlike several members of Diefenbaker's Cabinet, Pearson and his colleagues did not have an entrenched position on the still-unratified treaty. Soon talks were underway with the Americans as well. The Canadian side was led by Paul Martin, Pearson's Minister of External Affairs and a skilled negotiator, who worked closely with several senior BC cabinet ministers. The result of these discussions was a protocol—in effect, a revised treaty—signed on 22 January 1964 by Prime Minister Pearson and

President Lyndon Johnson, with the enthusiastic support of Bennett and his government. As the Victoria *Daily Colonist* observed, "Ottawa composed the music but Bennett wrote the words. It is a major political triumph for the Bennett government and particularly for the Premier."[30] As the quotation suggests, the treaty gave Bennett what he wanted. The province received cash for the downstream benefits, money that would be used to pay for the construction of the three treaty dams in British Columbia. Since one of these, the Mica Dam, would also generate considerable electricity, Bennett made his well-publicized claim that the province would obtain free power from the Columbia, "And nothing is freer than free, my friend."[31]

If the treaty was a triumph for Bennett — he spoke of it as his greatest achievement — others regarded it as a sell-out or worse.[32] A number of Canadians questioned the wisdom of exporting energy to the US market, regarding the relationship between the two countries as unequal and one in which the interests of the US would inevitably dominate. The rising tide of Canadian nationalism in the 1960s heightened such concern, which was also magnified by a widespread distaste for American foreign and domestic policies (notably with the Vietnam War abroad and racial conflict at home).[33] The Columbia River treaty was only one in a long series of issues during this period which focused attention on the nature of the Canada-US relationship. (Other highly controversial matters included the great pipeline debate of 1956 and the BOMARC missile crisis of 1962–1963.) In part, these reflected the growing integration of the Canadian and American economies in the postwar era, culminating finally in the Canada-US Free Trade Agreement of 1988. Economists and political scientists tended to see this continentalist thrust as the inevitable result of the logic of technology and trade. For nationalists of all political stripes — from the Conservative leader, John Diefenbaker, to members of the NDP "Waffle" — such an outcome could not be reconciled with their vision of an independent Canada, hence their opposition to all its manifestations.

Criticism of the Columbia River treaty was not only fuelled by nationalism, however. Others argued that its provisions for the sale of downstream benefits effectively denied the province a very real economic advantage. This was the reason that Dal Grauer, the head of BC Electric, opposed the Columbia River deal as well as Bennett's two river policy.

In the Kootenays, the treaty's impact was considerable, in the short as well as in the long term. The most obvious — and controversial — result was the flooding of the Arrow Valley. Richie Deane, an electrical engineer first with West Kootenay Power and then with Cominco, led a

> ### *The cheapest future power*
>
> The potential power in British Columbia is from a number of sources and will therefore be at differing costs. It would seem to be sound public policy to keep the cheapest power for British Columbian use and to stimulate economic development here, rather than to export the least expensive power. This point seems to have particular relevance because the cheapest future power seems undoubtedly to be the downstream benefits of the Columbia River, and there might be some pressure from the United States to buy Canada's share at the American generating plants where it is produced, rather than returning it to Canada. To succumb to such pressure would, in my opinion, be a tragedy because most of British Columbia, unlike the states of the Pacific Northwest, has never had the stimulus of really cheap power. Now that the wheel of fortune in this respect is at last spinning in our direction, we should make every effort to take advantage of it.
>
> — **A. E. DAL GRAUER**, "The Export of Electricity from Canada," 1961[34]

spirited attack against the High Arrow dam, the cause of this flooding. Deane presented a thoughtful critique to the House of Commons' External Affairs Committee when it considered the final version of the treaty in the spring of 1964.[35] By that time, however, the negotiations were effectively over: the federal government was unwilling to alter the terms of a document which had been so long in the making. Kootenay residents were left with no choice but to live with the treaty's consequences, although their views had rarely been taken into account during the lengthy process that culminated in the final agreement of 1964.

The Columbia River treaty and — more generally — the Bennett government's attitude towards hydro-electric projects changed the very nature of the energy industry in British Columbia. As late as 1945, for example, West Kootenay Power generated more electricity than BC Electric produced for the whole of the Lower Mainland.[36] With the creation of the BC Hydro and Power Authority in 1962, however, West Kootenay Power was dwarfed by its Vancouver-based competitor. BC Hydro united BC Electric, the Peace River Power Development Company, and the British Columbia Power Commission (the government-owned utility created in 1945, described in the last chapter) in one large province-wide company. It subsequently assumed control of the Columbia River project as well.

West Kootenay Power could not match the resources or the authority at BC Hydro's disposal, but the company continued to expand the scope of its operations during the 1950s and 1960s. Much of this expansion reflected decisions made by Cominco, notably with the construction of Waneta plant in the early 1950s. At the same time, Cominco also arranged to have a transmission line built to take power from the Kootenay River plants east across Kootenay Lake and into the East Kootenay mining community of Kimberley, home to the company's Sullivan mine.

The transmission line across the lake — the Kootenay Lake Span — was the logical culmination of the plans first formulated in 1897, to provide electrical energy to the mines of the region. The Sullivan mine was the foundation of Cominco's prosperity once the difficulties with zinc treatment were resolved and — as with the Rossland and Boundary mines in earlier periods — it required a reliable and cost-effective power supply. Up until the Second World War, a thermal plant adjacent to the Sullivan provided electricity but the subsequent need to expand and renovate this plant encouraged Cominco's engineers to re-consider the mine's power source. They concluded that the most economical strategy was to replace the aging thermal plant with a link into West Kootenay Power's transmission grid, something that West Kootenay Power's engineers had contemplated as early as 1919.[37] Electricity from the Kootenay River plants could be brought to the Kimberley area in several ways, either across Kootenay Lake or by a longer land route around it. Despite the engineering challenges — at the time of its construction, the two mile span across the lake was the world's longest — Cominco decided to build across the lake.[38]

The Kootenay Lake Span was the most dramatic example of a general trend in the 1950s and 1960s. Throughout the period, West Kootenay Power extended its lines and gradually began to tie in with other transmission systems. For example, although Cominco had been unable to sell the excess power from the Waneta plant in the mid 1950s, negotiations in the early 1960s with the Bonneville Power Administration led to an agreement to link the two systems, allowing energy to flow either way. Canada's National Energy Board approved this scheme — the equi-change agreement — since technically no power was exported; at the end of each year an equal amount of electricity flowed north and south. Earlier — in the late 1950s — West Kootenay Power built a more effective link in the western end of its operations, so that electricity could be purchased from the BC Power Commission, if necessary, for the Okanagan Valley. And in the East Kootenays, the company installed back-up switching gear so that Kimberley and the Sullivan Mine could obtain electricity from the East Kootenay Power Company should any difficulties interrupt transmission across the Kootenay Lake Span. The company's generating plants along the Kootenay River were also modernized during this period, following

The Kootenay Lake Span. Workers assemble the Span's East Tower in November 1951.

the practice of other power companies, with automation and central control replacing the close supervision of individual operating staff in each powerhouse.

Consumers — and would-be consumers — also became more sophisticated. People had come to expect access to a reliable electrical supply, and complained long and loud when it was not available. The files of the Public Utilities Commission contain numerous cases of people's heightened expectations and growing assertiveness. In the autumn of 1966, for example, the Commission held hearings in the Princeton area to consider the demand for electrical service in more

WEST KOOTENAY POWER – TRAIL OFFICE

Not a job for the faint-hearted: a West Kootenay Power employee works on the cable's vibration damper high above the shore of Kootenay Lake.

out-of-the-way parts of that region. The following was a hand-written brief submitted by the residents of Otter Creek, north of Tulameen, and is a good illustration of people's attitude to power by the 1960s.

> ### *Power is not a luxury*
>
> We, residents of Otter Valley, ranchers and summer people with homes, feel that we should be entitled to the conveniences of electricity.
>
> In this day and age power is not a luxury.
>
> There are school children and pre-school children in the valley, and why should they be deprived of good lights and have their eye sight ruined, when electricity is only a few miles away.
>
> Nearly all the permanent residents are ranchers. They need power for many uses outside. One use, especially, is for sprinkler irrigation, to improve their hay, grain crops and gardens. With more hay they can raise more beef. Approximately 500 or more acres could be under irrigation in the Valley.
>
> The labour of women in the Valley can be cut down with the use of electricity. Quite a few women work out-side and still have to struggle with the old methods of house-cleaning, washing, ironing and cooking.
>
> The house-hold budget can be cut down with the use of deep freezes and fridges.
>
> Where there are not any facilities to keep food, in many cases it has to be thrown away.
>
> — "A Brief on Power for Otter Valley"[39]

Representatives of West Kootenay Power attended the hearings and listened sympathetically. They tried to explain the company's objections to providing service to the Otter Lake area. It was an awkward situation for the company: as a private utility, it did not want to antagonize potential customers, nor could it ignore the guidelines and decisions of the Public Utilities Commission. Still, it could not escape the realities of the market, although the earnest explanations of company executives were likely not well-received.

People living in areas that did not enjoy access to power were well aware of the provincial government's support for large hydro-electric projects, which were very much before the public eye in the 1950s and the 1960s. They felt as if they were being ignored, that their more modest needs were overlooked amidst the inflated rhetoric of mega-projects. In Beaverdell, for example, a local representative contrasted all

Otter Lake Extension

The estimated cost of . . . the proposed Otter Lake extension would be $3,645 per year and the estimated revenue $750. Each customer would have to be subsidized $32 per month to provide for the cost of supplying service.

It seems unduly unfair that the revenue from customers we serve in other areas should be used to provide $192 each year to each summer home owner in the area around the north end of Otter lake. . . .

. . . West Kootenay Power has always been guided by a sense of responsibility to its existing customers, who I am sure do not want West Kootenay Power to engage in practices detrimental to the rates now enjoyed by them. Spending money on the construction of uneconomical extensions, and accepting the continuing burden of subsidies involved in the operation of such extensions, will hasten the need for the Company to seek an increase in rates. . . . We want to continue to serve our customers at low rates.

— J. D. MCMYNN on behalf of West Kootenay Power, 17 October 1966[40]

the excitement surrounding the Columbia and Peace River developments with the situation in his community:

> . . . we feel very strongly that this area needs power to grow. Now we have heard over the past ten years in this province a lot about power, that there is a gigantic power extravaganza, so to speak, occurring in the Province of British Columbia, and that we are developing to the tune of hundreds of millions of dollars the Peace and the Columbia River projects, and this is public funds which are being expended to finance these developments, and basically the reason behind it is, it is supposed to bring prosperity to this province, and when you ask, I think, people on the street what power means to them, they usually come up with the answer that power means industry. Now in this area we find it difficult to understand that almost the reverse is true. Here we are apparently told that if you want power you should have industry first, industry that is willing to contract in some cases for long-term power. . . .
>
> — DAVID JOHNSON, representing the people and businesses of Beaverdell[41]

Such appeals were essentially political in nature. (Not surprisingly, the residents of Beaverdell ensured that their local MLA attended the hearings and spoke up on their behalf.) The logic of the argument was a call for equitable treatment: if the provincial government could justify spending considerable amounts of public money on huge schemes, then Beaverdell deserved its share. However, West Kootenay Power could not afford to respond to political arguments, nor could it behave like a crown corporation. As its representatives were at pains to point out, the company had to keep its costs as low as possible, and it could not subsidize consumers in one area by raising the rates of consumers in other areas. Decisions about extending its operations had to take into account the company's overall profitability.

If most people in southeastern British Columbia now regarded power as essential, one group remained ambivalent to the apparent advantages of the modern age. The Doukhobors left Russia and came to Canada at the turn of the century to find freedom and escape persecution.[42] They soon found that slipping the bonds of oppression was rather more complicated than a simple ocean journey. Their deeply-held religious beliefs — a fundamental tenet of which was to deny any authority but that of their God — made them the object of mounting persecution in late nineteenth century Russia. In Canada, they were tolerated at first but suspicion and hostility soon surfaced. Many reasons could explain the ill will directed towards the Doukhobors — especially in the Kootenays — but perhaps the most compelling was simply ignorance and lack of communication. The new immigrants kept to themselves on their settlements; assimilation into Canadian

WEST KOOTENAY POWER – TRAIL OFFICE

West Kootenay Power employees pose for the camera in 1951, before heading off to work on the Kootenay Lake Span.

WEST KOOTENAY POWER – TRAIL OFFICE

West Kootenay Power's new transmission line into Kimberley crossed some rugged country; heavy snowfalls in the area meant that regular winter patrols were necessary.

society was not the reason that they had come to this adopted home. The Doukhobors' lifestyle reflected both their background and their belief system: largely rural people unaffected by modernist currents, they cultivated a strong sense of community and religious solidarity, and cherished a profound resistance to secular authority.

The Doukhobors also had a long-standing tradition of charismatic leadership. One such person — Peter Verigin — had led them to Canada at the end of the nineteenth century, organizing settlements on the prairies and in British Columbia. He was killed in a railway explosion in 1924, and evidence suggests that it was a deliberately planned execution.[43] The events surrounding his death reveal that the Doukhobors were not — and probably had never been — a monolithic group. Various sects rose and fell within their communities, although one stood out and achieved widespread notoriety. It was called the Sons of Freedom.[44] Members of the Sons of Freedom routinely destroyed property, an act which carried important religious significance, symbolizing a denunciation of worldliness and the profane secular life. When they burned down their homes or schools, or paraded nude, they attracted much critical attention, but apart from the bluster of newspaper editorials or the speeches of indignant Anglo-Canadian politicians, the consequences were not usually severe. However, the Sons of Freedom on occasion attacked public buildings or the property of others. The transmission lines of West Kootenay Power — just across the river from their community at Brilliant — became frequent targets of sabotage.

Company employees grew accustomed to finding vandalized poles on their patrols of the transmission lines. One person who worked for West Kootenay Power during this period recalled the many damaged poles that he had encountered:

Sabotage

. . . we had to get out and repair them, stub them. They would fill up an old bicycle tube with powder and wrap them around the pole, and light it. Some it would knock over; others it would just take a big chunk out of it . . . It started back in the 1940s and went all the way up to the 1950s . . . we got so we were kinda frightened going around the poles. When you were out patrolling, you didn't know if the damned thing was going to go off any time or not. . . . Some of them were timed with old alarm clocks and others were not, so you didn't know when they would go off. So you were always looking at the pole pretty close before you got up to it. . . .

— **MR. TATANGELO**, former West Kootenay Power employee

The sabotage was rarely effective — another person wryly noted that "They did not seem to be too expert at setting these dynamite charges"[45] — and as often as not it was more an inconvenience than a serious threat to the company's power grid. At worst, a pole was blown over and a transmission line temporarily knocked out. Naturally, the company did not welcome such attacks on its property. In 1951, the president complained in his Annual Report that "The company suffered from sabotage and malicious damage to transmission lines during July and August, resulting from blasting of poles with dynamite and damage to insulators by rifle fire. The perpetrators of these crimes were not apprehended."[46] However, far more significant damage was to occur a decade later, with potentially life-threatening consequences.

WEST KOOTENAY POWER – TRAIL OFFICE

March 1962: West Kootenay Power employees survey the collapsed East Tower. They erected a replacement span in record time.

The early 1960s marked a crescendo of attacks by the Sons of Freedom, the period when the group "embarked on the most dramatic and possibly the last of all their great campaigns against that "corrupt" North American culture that has both repelled and tempted them by its materialism." Over a hundred acts of burning and bombing occurred in 1961; the following year, an estimated two hundred and seventy-four such acts took place. The most dramatic act came on the 6th of March 1962, when dynamite brought down the 240-ton East Tower of the Kootenay Lake Span and cut off power to the East Kootenays. Those affected included miners at Cominco's Bluebell mine at Riondel as well as the Sullivan mine in Kimberley. These workers depended on electricity for the very air that they breathed while underground, and the explosion could easily have led to numerous fatalities. They reacted angrily when news of the power failure's cause reached them. With threats of vigilante action in the air, the RCMP rushed reinforcements to the Kootenays, arresting a number of Doukhobors on conspiracy charges. The atmosphere in Nelson became so tense that the trial had to be moved to the coast. Civil liberties groups were appalled by the government's heavy-handed response, while the government's case against those that had been arrested was so thin that the magistrate dismissed the case. Ironically, the subsequent "Great March" of the Sons of Freedom out to the coast, undertaken as a protest against the arrest and imprisonment of some of their members, helped to reconcile them to the larger society with which they had had such a tempestuous relationship.[47]

The bombings of the 1960s frequently tarred all Doukhobors with the notoriety earned by a small group of Freedomites. This was unfair. The Doukhobors had come to play a significant role in the region, and their lives were not unlike those of their neighbours, whatever differ-

An official ceremony marks the inauguration of cross-border power transmission in the 1960s. The first power exchanges were governed by the "equichange" agreement, which stipulated that the amount of power annually transmitted across the border had to balance, with equal amounts going each way.

ences they may have had with the dominant values of BC society. And at times even these differences had unlooked-for benefits. For example, their strong pacifist beliefs meant that Doukhobor men were exempted from military service during the Second World War. A general labour shortage during the war as well as the proximity of their settlements to the Brilliant dam made them obvious candidates to work on the building project. Hundreds of Doukhobor men helped with the dam's construction.

West Kootenay Power's staff worked at remarkable speed to restore the Kootenay Lake Span. They succeeded in putting a temporary line into operation less than a month after the destruction of the original span. People who took part in the repair job had a real sense of pride in their accomplishment:

> Dominion Bridge said it would take at least two months, sixty days, to reconstruct it and even at that they could not guarantee that they could have it done at that time. With all Cominco's operations in the East Kootenays plus Creston and Riondel . . . out of power at that time, it was critical that the power be restored at the earliest possible date. Art Lambert called Jimmy McClelland, Bill Robson and myself into his office and asked us what we thought about constructing a temporary span across there. . . we came out with the unanimous answer that we thought we could do it and we were given the green light, although Dominion engineering said that it was impossible. . . . It was a challenge and the hours of the day didn't mean much to us. . . . To see the thing take shape and go across in that period of time was really something . . . it was a tremendous effort and it was accomplished in spite of all the leading engineering brains at the time saying it could not be done. West Kootenay Power and Light Company crew did it.
>
> — STU METCALFE, former West Kootenay Power employee

> The new ⅝ in. cable span was completed by April 2, just 28 days from the date of the blasting of the main span. This was undoubtedly a remarkable feat of design, procurement, manufacture and construction under most difficult conditions as March winds are not friendly and the West Tower site was covered by 3 ft. of snow.
>
> — R. DEANE, former Chief Electrical Engineer, Cominco[48]

The repair job was not intended to be permanent. In the following months, Dominion Bridge crews put in a new span, which incorporated several improvements over the original 1952 design, reflecting the experience of the intervening ten years as well as changes in Department of Transport regulations concerning the minimum clearances permissible between the lake and the span's lowest point.[49]

Changes during the 1960s — the Columbia River treaty, the formation of BC Hydro, the continuing growth in energy consumption — all had an impact on West Kootenay Power. Reflecting on these developments in an inter-office memo entitled "Legal Considerations: Cominco and West Kootenay Power," Cominco's lawyer noted that they "have added a new dimension to the electric power potential of Cominco and West Kootenay Power." The memo went to West Kootenay Power's general manager. Joe Drennan read the four page document carefully, highlighting the quoted passage. A man of few words, Drennan wrote in response: "Not sure what a new dimension is. Has not increased our power supply. Has made for a more complex, integrated power operation."[50] The company's transmission lines stretching west, south, and east — where they joined those of BC Hydro, the Bonneville Power Administration, and Calgary Power — reflected what Drennan had in mind.

Affluence and growth, the hallmarks of the post-war era, left consumers with an insatiable appetite for energy. More and more power flowed in every direction along the company's transmission lines. The arrangement between West Kootenay Power and its parent company was re-defined several times, in an effort to acccommodate the growing complexities that Drennan identified. By the 1980s, however, the challenge was too much. Government regulation and a volatile economic climate combined to change fundamentally the relationship between the two companies.

NOTES

1 These figures are from Mary Sue Devereaux, "New Necessities: Popular Household Appliances," *Canadian Social Trends*, #10 (Autumn 1988): 31-34. Much of the data for this article comes from Statistics Canada's Catalogue 64-202, a fascinating source for tracking Canadians' growing use of household appliances. (Data has been collected annually since 1948.) For more recent statistics which chronicle the rapid adoption of VCRs, CD players, and home computers over the past decade and more, see Jeffrey Frank, "Preparing for the Information Highway: Information Technology in Canadian Households," *Canadian Social Trends*, #38 (Autumn 1995): 2-7.

2 For a similar analysis, see Donald V. Smiley, "Canada's Poujadists: A New Look at Social Credit," *Canadian Forum*, Vol. 42, #500 (September 1962): 121-23.

3 The following discussion is drawn from a number of sources, but I am particularly indebted to the comprehensive account provided by Neil A. Swainson, *Conflict over the Columbia: The Canadian Background to an Historic Treaty*, Montreal: McGill-Queen's University Press, 1979.

4 These figures are from J. D. McDonald, *Storm over High Arrow: The Columbia River Treaty (A History)*, Rossland: privately printed, 1993, 3. See also John V. Krutilla, *The Columbia River Treaty: The Economics of an International River Basin Development*, Baltimore: Johns Hopkins Press, 1967, 15-22.

5 A. A. Lambert and Gordon Dunbar, "Hydro-Electric Development at Lower Bonnington Falls on the Kootenay River," typescript, 1964, 6 (copy in West Kootenay Power's Trail office).

6 See Paddy Sherman, *Bennett*, Toronto: McClelland and Stewart, 1966, 210; also, Swainson, *Conflict over the Columbia*, 42.

7 See the discussion in L. M. Bloomfield and Gerald F. Fitzgerald, *Boundary Waters Problems of Canada and the United States (The International Joint Commission 1912–1958)*, Toronto: The Carswell Company, Ltd., 1958, 190-95, and Swainson, *Conflict over the Columbia*, 45-51.

8 See Bloomfield and Fitzgerald, *Boundary Waters Problems*, 196, and Swainson, *Conflict over the Columbia*, 48-49.

9 I am indebted to Jack McDonald for this information. See also Interview #3 (Stuart Metcalfe), 6.

10 See E. M. Coleman, "West Kootenay Power," typescript draft, 1976, 4-5 (copy in West Kootenay Power's Trail office).

11 A. E. Dal Grauer, "The Export of Electricity from Canada," in R. M. Clark, ed., *Canadian Issues: Essays in Honour of Henry F. Angus*, Toronto: University of Toronto Press, 1961, 279; Interview #14, Art Lambert, 6; and interview with Richie Deane, 2 February 1997.

12 See Swainson, *Conflict over the Columbia*, 57-65, and Sherman, *Bennett*, 211-14. David Mitchell incorrectly asserts that the Kaiser proposal concerned the Mica dam, north of Revelstoke (*W. A. C. Bennett and the Rise of British Columbia*, Vancouver: Douglas & McIntyre, 1983, 285).

13 I have taken this date from Sherman's account in *Bennett*, 214-15; Martin Robin, on the other hand, has the meeting take place two years earlier: see his *Pillars of Profit: The Company Province, 1934–1972*, Toronto: McClelland and Stewart, 1973, 207-208.

14 On the Wenner-Gren talks, their outcome, and the subsequent controversy, see John R. Wedley, "Infrastructure and Resources: Governments and their Promotion of Northern Development in British Columbia, 1945–1975," PhD thesis, University of Western Ontario, 1986, 247-310. Other useful accounts include Sherman, *Bennett*, 209-30, and Robin, *Pillars of Profit*, 207-12.

15 In fact, the federal government felt that the Peace River project required a federal license, although BC Hydro never bothered to apply for one; see Swainson, *Conflict over the Columbia*, 217-18, 404.

16 This analysis is based largely on detail provided by Swainson, *Conflict over the Columbia*, passim, and Wedley, "Infrastructure and Resources," 247-310. Useful contemporary accounts include Stuart Keate, "The Smile of the Tiger: Why Bennett Took Over BC Power," *Saturday Night*, Vol. 76 (16 September 1961): 11-14, and John A. Irving, "Bennett's Design for BC Progress," *Saturday Night*, Vol. 77 (26 May 1962): 18-24, while the premier's somewhat expansive memories of these events may be gleaned from Mitchell, *W. A. C. Bennett and the Rise of British Columbia*, 286-327.

17 For an indication of his private anxiety during this period, however, see Mitchell's *W. A. C. Bennett and the Rise of British Columbia*, 306-15.

18 M. H. Mason, Cominco's general counsel in 1976, was present at the special session of the Legislature, when the bill was brought in to nationalize BC Electric. Mason recalled

that "In introducing the Bill he [Bennett] remarked that West Kootenay Power was next on the list." ("Legal Considerations: Cominco and West Kootenay Power," typescript, 1976, 3).

[19] This view was common among senior management within the company: see for example, the comments of Doug McMynn, Interview #17, 6, and M. H. Mason, "Legal Considerations: Cominco and West Kootenay Power," typescript, 1976, 3-4 (copy in West Kootenay Power's Trail office). Similarly Jack McDonald — an electrical engineer with a lifetime's experience observing the company's career — repeated this analysis in conversation with me.

[20] See also the following accounts: Robin, *Pillars of Profit*, 230-32; Mitchell, *W. A. C. Bennett and the Rise of British Columbia*, 303-05; Stuart Jamieson, "Power in BC," *Canadian Forum*, Vol. 41, #496 (May 1962): 35-36; Keate, "The Smile of the Tiger"; Irving, "Bennett's Design for BC Progress."

[21] Quoted in Sherman, *Bennett*, 245. Bennett was right about the projects' size, however, which was such that leading American officials seriously doubted that money could be found to finance them both at the same time (Swainson, *Conflict over the Columbia*, 262).

[22] Quoted in Mitchell, *W. A. C. Bennett and the Rise of British Columbia*, 305.

[23] Mitchell, *W. A. C. Bennett and the Rise of British Columbia*, 307.

[24] This issue is contentious; see Swainson's comments, *Conflict over the Columbia*, 180-85, and the contemporary account by Bruce Hutchison, "The Great Columbia River Foul-Up," *MacLean's*, Vol. 74, No. 11 (3 June 1961): 15, 60, 62-63.

[25] The Vancouver publisher, Howard Mitchell, quoted in Sherman, *Bennett*, 242.

[26] See Grauer, "The Export of Electricity from Canada," 248-85; Sherman, *Bennett*, 245-46; and James Robert Allum, "Science, Government and Politics in the Abolition of the Commission of Conservation, 1909–1921," MA thesis, Trent University, 1988, 76-81.

[27] See Swainson, *Conflict over the Columbia*, 232; Ronald B. Worley, *The Wonderful World of W. A. C. Bennett*, Toronto: McClelland and Stewart, 1971, 243; and Sherman, *Bennett*, 260-62.

[28] See Peter C. Newman, "Backstage in Ottawa: Who's where and why in the great Columbia River Debate," *MacLean's*, Vol. 75, No. 23 (17 November 1962): 2-3; note also the comments in Robin, *Pillars of Profit*, 239-40, and Sherman, *Bennett*, 260. Swainson provides contrary evidence, although his source is scarcely an impartial one: *Conflict over the Columbia*, 316.

[29] Sherman, *Bennett*, 272. The 1963 provincial election brought into sharp focus the issues surrounding the nationalization of BC Electric and the controversy over the Columbia River Treaty. For accounts of the campaign, see Sherman, *op. cit.*, 271-77; Robin, *Pillars of Profit*, 240-49; Mitchell, *W. A. C. Bennett and the Rise of British Columbia*, 317-21; and Swainson, *Conflict over the Columbia*, 259-63.

[30] Quoted in Worley, *The Wonderful World of W. A. C. Bennett*, 244.

[31] Quoted in Mitchell, *W. A. C. Bennett and the Rise of British Columbia*, 324.

[32] On Bennett's pride in the Columbia treaty — the 1964 version — see the comments in Mitchell, *W. A. C. Bennett and the Rise of British Columbia*, 325, and Sherman, *Bennett*, 281. For a very different analysis, from two of the treaty's harshest critics, see Larratt Higgins, "The Alienation of Canadian Resources; The Case of the Columbia River

Treaty," in Ian Lumsden, ed., *Close the 49th Parallel Etc.*, Toronto: University of Toronto Press, 1973, 223-40, and General A. G. L. McNaughton, "The Proposed Columbia River Treaty," *International Journal*, Vol. 18, No. 2 (Spring 1963): 148-65.

33 Contemporary Canadian views of the US may be gleaned from Al Purdy, ed., *The New Romans: Candid Canadian Opinions of the US*, Edmonton: M. G. Hurtig, 1968.

34 Grauer, "The Export of Electricity from Canada," 283. In a haunting irony, Grauer was buried the same summer's afternoon in 1961 as the act nationalizing his company was introduced in the Legislature.

35 Richie Deane, "The Columbia Treaty and High Arrow Dam," typescript submission to the External Affairs Committee of the House of Commons, 30 March 1964. I am grateful to Mr. Deane for kindly supplying me with a copy of his submission. For evidence of the effectiveness of Deane's submission, see Swainson's comments, *Conflict over the Columbia*, 275. For the protests and views of others in the Kootenays who objected to the treaty's provision of the High Arrow dam, see Donald Cresswell Waterfield, *Continental Waterboy: The Columbia River Controversy*, Toronto: Clarke, Irwin, 1970, and J. D. McDonald, *Storm over High Arrow: The Columbia River Treaty (A History)*, Rossland: privately printed, 1993.

36 See the statistics in British Columbia, *Report of the Rural Electrification Committee As of January, 1945*, Victoria: King's Printer, 1945, 12, 14. These show that BC Electric's Lower Mainland stations produced 231,455 KVA, while West Kootenay Power's five Kootenay River plants produced 260,500 KVA. An internal company report later claimed that "During the early 1940s the Zinc Tank Rooms in Trail used more electricity than did the City of Vancouver." (Lambert and Dunbar, "Hydro-Electric Development at Lower Bonnington Falls on the Kootenay River," 5).

37 E. M. Stiles, "The Kootenay Lake Span," *Engineering Journal*, Vol. 35, #8 (August 1952), 815.

38 This became the topic of several articles: E. M. Stiles, "The Kootenay Lake Span," *Engineering Journal*, Vol. 35, #8 (August 1952): 815-19, and E. M. Stiles, "The Kootenay Lake 2-Mile Transmission Line Crossing," paper presented to the Conférence Internationale des Grands Réseaux Électriques à Haute Tension, Paris, 1954 (a copy of this paper survives in the Cominco Papers, Add Mss 2500, Box 422, file 10, BCARS). Cominco contracted out the construction of the span itself to the Dominion Bridge Company.

39 I have corrected several spelling mistakes in the original, to avoid sprinkling the quotation with [*sic*]. The hand-written brief has no date or signature, but the next item is a brief dated 17 October 1966 entitled "Otter Lake Extension," and another by M. H. Mason, Counsel for West Kootenay Power, entitled "Statement on behalf of West Kootenay Power and Light Company, Limited to the Public Utilities Commission Hearing at Tulameen on October 17th, 1966." It seems safe to assume that the petition was presented at this Tulameen hearing, and on that date. The material is held in the papers of the British Columbia Energy Commission 1929–1979, GR1390, Box 27, BCARS.

40 Typescript held in GR1390, Box 27, BCARS. The brief prepared by M. H. Mason, Counsel for West Kootenay Power, "Statement on behalf of West Kootenay Power and Light Company, Limited to the Public Utilities Commission Hearing at Tulameen on October 17th, 1966," offers a very similar analysis.

[41] British Columbia, Public Utilities Commission, "In the matter of the "Public Utilities Act" and In the matter of a requested extension of the West Kootenay Power & Light Company Limited's Transmission and Distribution System into the Beaverdell and Carmi Areas," typescript of hearing at Beaverdell, BC, 18 October 1966, 8-9; copy in the papers of the British Columbia Energy Commission 1929–1979, GR1390, Box 27, BCARS.

[42] The following comments are drawn largely from George Woodcock and Ivan Avakumovic, *The Doukhobors*, Toronto: Oxford University Press, 1968.

[43] See Woodcock and Avakumovic, *The Doukhobors*, 257-59.

[44] For a description and analysis of this controversial group, see Woodcock and Avakumovic, *The Doukhobors*, 193-96, 308-31.

[45] Grant Tindale, Interview #18, 10; similarly, R. G. Anderson, a former general manager and president of the company, commented that "They were never able to do much damage. . ." (Interview #19, 8). The other quotation in the text, by Mr. Tatangelo, is from Interview #12, 9.

[46] R. W. Diamond, in West Kootenay Power and Light Company, Ltd., 1951 Annual Report, 2.

[47] This paragraph relies heavily on the analysis provided by Woodcock and Avakumovic, *The Doukhobors*, 350-55; the quotation in the text is from 350.

[48] Stu Metcalfe, Interview #3, 8-10; R. Deane, "Redesign of the Two Mile Kootenay Lake Power Span," *Engineering Journal*, Vol. 47, #6 (June 1964), 33.

[49] These changes are described in Richie Deane's article, "Redesign of the Two Mile Kootenay Lake Power Span."

[50] M. H. Mason, General Counsel, Cominco, "Legal Considerations: Cominco and West Kootenay Power," typescript, 1976, 4; cover memo, Vice-President & General Manager, West Kootenay Power (Joe Drennan), to General Counsel, Cominco, 11 August 1976, copy in West Kootenay Power's Trail office. Mason was also West Kootenay Power's lawyer.

CHAPTER EIGHT

The Eighties and After

In 1987 Cominco sold West Kootenay Power amid considerable controversy, seventy-one years after purchasing it from the Canadian Pacific Railway. The sale to an American company attracted much attention: politicians, journalists, as well as many local citizens participated in what became a heated and emotional debate about foreign ownership. Notwithstanding the blaze of publicity, the change in ownership was in many ways the culmination of events that dated back to the Second World War. Close regulation of the electrical industry, in particular, encouraged Cominco to get out of the utility business.

Provincial governments formed public utilities commissions in order to regulate such essential services as urban transportation networks, telephone systems, and domestic electrical supply. The growing importance of these services raised concerns about their cost and their availability, and the commissions were to ensure that utility companies did not take advantage of their customers. In British Columbia, the Public Utilities Commission's role as a public watchdog meant that West Kootenay Power needed its approval for any rate increase. The Commission granted such requests only after it was satisfied that the price hike was justified and necessary. This may not seem a very significant issue; after all, government regulation of utilities does not often capture people's attention. However, the lengthy and convoluted hearings that followed West Kootenay Power's application for a rate increase in 1976 led directly to Cominco's decision to sell the company. In fact, Cominco and West Kootenay Power were involved in two sales in the 1980s, although the first of these transactions attracted little publicity.

As early as the mid 1960s, the Public Utilities Commission expressed concern about the position of West Kootenay Power. Although the company provided its customers with relatively inexpensive electricity, the Commission questioned the nature of its relationship to Cominco. The problem lay in the fact that Cominco was the sole shareholder of West Kootenay Power. The Commission found it hard to believe that negotiations between the two companies — to determine the price at which West Kootenay Power would purchase electricity from Cominco, and thus the price that it would charge its customers — would be conducted in an open and impartial manner, or that these negotiations would necessarily arrive at a fair market price. The chairman of the Commission wrote to the president of West Kootenay Power, hoping to put the matter to rest.

The Commission considers . . .

The Commission has considered very carefully the evidence produced by West Kootenay Power and Light Co. Ltd. to show that the terms on which it is at present buying energy from its parent company are fair and reasonable in the sense that it is not paying more for energy than a fair market price. This evidence is of great importance because if the Commission were satisfied that the price which consumers are paying for power purchased from West Kootenay Power and Light Company Limited were not an inescapable cost to West Kootenay Power and Light Company Limited there would be no need for a general rate case on this matter. . . .

The Commission considers that West Kootenay Power and Light Co. Ltd. has not exhausted the possibility of obtaining a better price from Cominco, as it is open to West Kootenay Power and Light Co. Ltd. to apply to the Public Utilities Commission to deal with Cominco as a public utility and to regulate its rates. . . .

If it is impossible to deal with the price of power on the basis of Cominco's costs on a public utility basis the Commission may be compelled to raise the question of whether a public utility with a single shareholder is entitled to as high a rate of return as it would be allowed if its shareholder were not receiving benefits in addition to dividends. . . .

— **H. F. ANGUS**, 7 January 1965[1]

An exchange of letters followed, and the Commission was never completely satisfied with the answers it received.[2]

The issue re-surfaced ten years later, when West Kootenay Power was before the Commission once again. Things were a little different now: in 1972, Dave Barrett and the New Democratic Party replaced W. A. C. Bennett and the Socreds in Victoria. The following year the new government passed its "Energy Act," legislation which — among other things — replaced the Public Utilities Commission with the British Columbia Energy Commission.[3] This involved more than simply a change in the Commission's name, as a West Kootenay Power memo noted in the late 1970s: "The members of the [British Columbia Energy] Commission and staff were all new appointments. . . . They attempted to make their public hearings an arena for debate and they supported the antagonistic groups, such as environmental, consumer associations and others that have appeared in recent years."[4] These groups, however, were not the cause of what was soon to become a real problem for Cominco and West Kootenay Power. The difficulty lay in a seemingly unimportant clause of the new act, which stated that corporations generating power for their own use would not be regulated as utilities, as long as they did not sell more than fifteen per cent of their power to other users.[5] Presumably if they *did* sell more than this percentage, they would be regulated as a utility. This raised once again the question of whether Cominco ought to be declared a public utility — and regulated as one — since a growing amount of the energy it produced went to other users.[6] This had not always been the case, of course. From the First World War until the 1950s, Cominco absorbed almost all of the output of West Kootenay Power. By the 1970s, however, domestic consumers had become more significant and accounted for an ever-increasing proportion of the power produced by the plants along the Kootenay River.

The 1970s were years of runaway inflation, something that affected everyone in one way or another. As interest rates soared, businesses had to ensure that shareholders received a return on their investment comparable to what was available elsewhere in the money markets. This helps to explain why West Kootenay Power applied to the Commission in early 1976 for a rate increase, seeking to raise the price it charged customers by 12.6%. Although hearings dragged on intermittently for more than four years, the Commission granted the company's request for an increase on an interim basis in December 1976.[7] That, however, was the easy part. As the Commission noted in its decision, the central issue was "the cost of power and whether the price paid by West Kootenay to Cominco in 1976 was reasonable." The Commission wanted to discover how West Kootenay Power purchased energy from Cominco, an issue that raised serious questions about the efficacy of utility regulation.

> ### *Cost of Power*
>
> In a non-arms-length relationship such as exists in this case, a contract does not in itself automatically justify as appropriate, the price or cost of power between the parties.
>
> The significant fact is that West Kootenay is a subsidiary of Cominco and in such a clearly non-arms-length relationship all transactions between the parent and its regulated subsidiary must be thoroughly examined. In this case it must be recognized that West Kootenay should pay no more for the power it buys from Cominco than it would cost the utility if it owned and operated the facilities involved. Were it otherwise, the entire function of utility regulation would be jeopardized, as the means would readily be at hand to remove any utility from effective public scrutiny and control.
>
> — **BRITISH COLUMBIA ENERGY COMMISSION**, 1980[8]

The last thing that Cominco wanted was to become a regulated public utility; the company concluded that the situation called for drastic action. In January 1980, West Kootenay Power sought the Commission's approval to lease from Cominco — with an option to buy — Plants 2, 3, and 4 (Upper Bonnington, South Slocan, and Corra Linn). These were the plants that Cominco had acquired in 1947, shortly after the death of Lorne Campbell. There was a catch: if the deal went ahead, Cominco wanted the provincial government to exempt it from the provisions of the Energy Act.[9]

The BC Energy Commission accepted the proposal but the Cabinet apparently rejected the idea and so it was shelved.[10] Bill Bennett and the Social Credit party had taken over the reins of government in 1975 and they subsequently replaced the NDP's Energy Act with their own legislation — the Utilities Commission Act — which altered utility regulation in the province once again.[11] In early 1981, Cominco and West Kootenay Power re-applied to the new BC Utilities Commission, seeking approval of the sale of the three Cominco plants back to their original owner. In addition, Cominco sought exemption from regulation as a utility and agreed to finance the sale of the plants to West Kootenay Power.[12] The Commission duly forwarded this second proposal to the Cabinet for its consideration. This time, the government approved the deal.[13]

Announcing the Cabinet's approval of the sale as well as Cominco's exemption from regulation as a utility, Energy Minister Bob McClelland referred to the deteriorating financial climate and the rather bleak position in which Cominco found itself.[14] The early 1980s were certainly not good years for the company and the re-arrangement with West Kootenay Power was only one part of an ongoing process of change and adjustment. Confronting low metal prices and ballooning debt, Cominco began to sell off assets that were not absolutely crucial to its core business of mining and smelting. In 1981 it sold Pacific Coast Terminals; four more subsidiaries were up for sale in 1986, a year in which Cominco's net loss was a little over a hundred and fifty million dollars. One of these companies was West Kootenay Power.[15]

Cominco's sale of three plants to West Kootenay Power in 1982 attracted very little comment. This was certainly not the case in August 1986 when the company announced that it had accepted an eighty million dollar bid from the Missouri-based UtiliCorp for West Kootenay Power. Eleven months of bitter controversy followed, ending only when the deal cleared the last regulatory hurdle — the approval of the BC Utilities Commission — in July 1987.[16]

From Cominco's perspective, the deal was simply that, a deal. UtiliCorp's bid was substantially higher than the others that it had received, and the whole point of the sale was to raise money for the cash-strapped company. Ironically, the controversy over the UtiliCorp purchase coincided with the sale of Cominco itself. In the autumn of 1986, Vancouver-based Teck Corporation, with help from corporate partners in Germany and Australia, bought the company from CP.[17]

UtiliCorp was the name adopted by the Missouri Public Service Company in 1985. The change was a symbolic one, signalling the company's desire to grow beyond its home region and to grasp advantages in the energy business wherever they could be found. The acquisition of West Kootenay Power was consistent with a larger vision, and also reflected a shift already underway in the US to de-regulate the power industry. (The company is now active in eight states as well as British Columbia, and also has operating divisions in New Zealand and the United Kingdom.)[18]

Many residents of the Kootenays opposed the sale, arguing that West Kootenay Power should remain Canadian-owned. There was also concern that the sale would bring with it substantially higher rates. Much of the controversy, however, reflected the ways in which this was a local chapter in a larger national debate over the nature and direction of the relationship between Canada and the United States. The election of Brian Mulroney and the Conservatives in 1984 served to bring the issue into sharp focus.

Not long after the Conservatives took office in Ottawa, Prime Minister Mulroney announced his government's plan to negotiate a free trade agreement with the United States. The abolition of federal boards such as the Foreign Investment Review Agency, combined with Mulroney's personal friendship with US President Ronald Reagan and his employment history as the senior executive of a large US firm operating in Canada, further antagonized Canadian nationalists. In this charged atmosphere, the sale of West Kootenay Power was portrayed as an example of betrayal and worse, in much the same way that the Columbia River Treaty had been a lightning rod twenty years earlier for those who championed Canadian ownership and control of the country's natural resources. Despite the controversy, neither federal nor provincial regulatory bodies opposed the sale, although the BC Utilities Commission did attach a number of conditions when approving the transaction in the summer of 1987.[19]

Whatever else it did, the change in ownership invigorated West Kootenay Power. Cominco had had little interest in the utility business and never invested in the company's future. The BC Utilities Commission mentioned this in a roundabout way in 1982: "there has been a general reluctance on behalf of the sole shareholder [in West Kootenay Power] to provide additional equity over a long period of time."[20] The new owner, on the other hand, has been anxious to demonstrate its willingness to improve and upgrade service, as well as to participate in the social life of the communities that it serves.

Change continues in the Kootenays, often as a consequence of events elsewhere but occasionally as a result of local initiative and action. Two very different currents will likely affect West Kootenay Power in the years ahead. Ongoing de-regulation of the energy industry will probably exert the greatest influence on the company. De-regulation dates from the early 1980s, when conservative governments in both Britain and the United States — under Margaret Thatcher and Ronald Reagan — embarked on controversial programs aimed at reducing the role of government. Canadians followed these developments closely, with a number of business people claiming that these were necessary reforms for Canada as well. After the 1984 election brought the Conservatives to power, Ottawa began to implement changes similar to those that had already taken place south of the border. (The telephone and airline industries are two conspicuous examples of this process.[21]) Although the energy industry is a provincial jurisdiction, it too faces deregulation, largely as a result of Victoria's desire to see BC Hydro operate on a more competitive basis.[22] The impact of this on West Kootenay Power is hard to gauge although it is certain to change the ways in which the company carries out its business.

Another recent development has its roots in the region itself. The origins of the Columbia Basin Trust can be traced to local residents' concern over the environmental impact of the Columbia River Treaty dams in the Kootenays, particularly the impact of the Duncan Dam on the fish habitat in Kootenay Lake.[23] (No environmental impact assessments had ever been done on the dams.) People also felt that the provincial government owed the region a portion of the windfall profits from hydro-electric developments in the area. The rivers of the Kootenays were generating a considerable percentage of BC Hydro's energy by the late 1980s, which in turn generated revenue for the provincial government's coffers. Very little of this money found its way back to the Kootenays, a region suffering from high unemployment and with a relatively poor economic outlook. In 1989 the Association of Kootenay-Boundary Municipalities took the initiative and created the Columbia River Treaty Committee. This committee engaged in extensive public consultation, staging two symposia — in 1993 and 1994 — to help formulate its proposal for a regional equity package. It subsequently approached the provincial government, seeking some form of settlement for the considerable social and environmental impact of the Treaty dams on the region. The government was willing to negotiate and in 1995 it enacted legislation creating the Columbia Basin Trust, although the Trust's structure and purpose were largely determined by the Columbia River Treaty Committee's 1994 symposium in Cranbrook.[24] A separate contract with the provincial government guaranteed stable funding for the Trust. The Trust used a portion of this money to purchase the Brilliant Dam from Cominco in 1996, and is contemplating further investment in expansion and power generation in the area.

Columbia Basin Trust Becomes Law

. . . The trust will work to ensure that regional values and priorities are reflected in the management of the more than $1-billion regional investment in energy infrastructure and other economic development projects to be made in the region over the next 10 years.

"These investments will provide tremendous economic opportunity to the people of the Columbia Basin region who so unjustly bore the brunt of the Columbia River Treaty 30 years ago," said [Energy Minister] Edwards.

Glen Clark, minister of employment and investment, said proposed energy projects at three existing dams (Keenleyside, Waneta and Brilliant) will provide more than 4,400 person years of direct construction employment — and considerable opportunity for skills training and business development.

— **MINISTRY OF EMPLOYMENT AND INVESTMENT** news release, July 6 1995

Contemporary developments such as the energy industry's deregulation and the emergence of the Columbia Basin Trust prove that West Kootenay Power and the region it serves will continue to make history, to change and adapt as necessary. The rapid advances in computer technology and the increasing pace of globalization, certainly, will alter the relationship with the world beyond the Kootenays. Much has changed in the hundred years since Sir Charles Ross and others began to plan the first hydro-electric development at Bonnington Falls, and no doubt much will change in the years ahead, yet the future is not likely to lessen people's dependence on the energy generated by the rivers flowing through the region. Power, as Matthew Boulton observed so long ago, is still what all the world wants.

NOTES

1 H. F. Angus, Chairman, Public Utilities Commission, to R. G. Anderson, President, West Kootenay Power & Light, 7 January 1965, in the papers of the British Columbia Energy Commission 1929–1979, GR1390, Box 27, File #17, U-8-1, West Kootenay Power & Light Co. Ltd., BCARS.

2 In addition to quoted letter, Angus to R. G. Anderson, 7 January 1965, see R. G. Anderson to H. F. Angus, 17 February 1965; H. F. Angus to R. G. Anderson, 4 March 1965; R. G. Anderson to H. F. Angus, 2 April 1965, all in GR1390, Box 27, File #17, U-8-1, BCARS.

3 "Energy Act," Chap. 29, in *Statutes of British Columbia, 1972–1973*, Victoria: Queen's Printer, 1973, 351-98.

4 "Report on the situation with the BCEC [British Columbia Energy Commission]," internal West Kootenay Power memo held in the company's Trail office. The memo is undated but evidence in the text indicates that it must have been written between 1977 and 1980.

5 "Where a corporation generates electric power primarily for its own industrial purposes, that corporation is not subject to the provisions of this Act in respect of electric service furnished to others if . . . the service . . . does not use more than fifteen per cent of the electric power generated by it." ("Energy Act," *Statutes of British Columbia, 1972–1973*, 357.)

6 Throughout this period, Cominco argued that it did not sell over fifteen per cent of its power but it had trouble persuading the Commission that this was the case. I am grateful to Bob Stone, Cominco's Vice President Finance and Chief Financial Officer, for describing this situation to me in a telephone interview.

7 The Commission's "Summary of Decision" noted that the 12.6% increase would give West Kootenay Power "the opportunity to earn approximately 10% on rate base and 15.78% on common shareholders equity. The Commission finds these tariffs and resulting rates of return to be fair and reasonable in the circumstances . . ." (British Columbia Energy Commission, "In the Matter of the Energy Act and In the Matter of an Application by West Kootenay Power and Light Company, Limited," Victoria: British Columbia Energy Commission, 1980). Two years later the Utilities Commission

offered a similar analysis when it approved a further increase: see the comments in British Columbia Utilities Commission, "In the Matter of the Utilities Commission Act, SBC 1980, c. 60 as amended And In the Matter of Application by West Kootenay Power and Light Company, Limited for amendments to filed rate schedules," Victoria: British Columbia Utilities Commission, 1982, 13, 23. As the Utilities Commission noted, three interim increases in 1979, 1980, and 1981 respectively raised the 1979 base rate a total of 55%, an indication of the severity of inflation by the end of the 1970s (*op. cit.*, 4).

8 British Columbia Energy Commission, "In the Matter of the Energy Act . . . ," 4. The quotation in the text is from the same source.

9 See British Columbia Energy Commission, "In the Matter of the Energy Act . . . ," 3, 8, and "West Kootenay rate hearing adjourns," *The Province*, 8 January 1980.

10 See the comments in "Cominco, Victoria in deadly power duel: Firm's sale of electricity runs afoul of utilities law," *Vancouver Sun*, 2 March 1981, F1.

11 "Utilities Commission Act," Chap. 60, in British Columbia, *Statutes of British Columbia, 1980*, Victoria: Queen's Printer, 1980, 351-92. The act came into effect on 11 September 1980.

12 See British Columbia Utilities Commission, "In the Matter of the Utilities Commission Act, SBC 1980, c. 60 as amended And In the Matter of Application by West Kootenay Power and Light Company, Limited for amendments to filed rate schedules," Victoria: British Columbia Utilities Commission, 1982, 5-6, and "Power deal 'money-saver'," *Vancouver Sun*, 13 August 1981, A14.

13 See "Cominco hydro sale again before cabinet," *Vancouver Sun*, 6 May 1982, D3; "Power plants change hands," *Vancouver Sun*, 29 July 1982, C5; "Three more industrial turbines added to the utility ledger," *Energy Analects*, Vol. 11 (6 August 1982): 3. For the formal arrangements, see British Columbia Utilities Commission, "In the Matter of the Utilities Commission Act . . . ," 6-10, 22.

14 For these comments, see "Power plants change hands," *Vancouver Sun*, 29 July 1982, C5.

15 See "Cominco selling Pacific Coast," *Northern Miner*, Vol. 67, No. 3 (26 March 1981): 30, and "Cominco set to sell off three assets," *Globe and Mail*, 19 June 1986, B6. In fact, Cominco put four significant assets up for sale in 1986. The three described in the headline were West Kootenay Power, Western Canada Steel, and the Con gold mine in Yellowknife, but earlier that year the company had sold its 40% stake in Fording Coal for $87 million. On Cominco's decision to sell West Kootenay Power, see "Cominco set to sell unit," *Globe and Mail*, 19 June 1986, B3; also "Cominco sells hydro utility to US firm," *Globe and Mail*, 19 August 1986, B3. The figure for Cominco's net loss in 1986 is from "Angry protests greet sale of BC utility," *Toronto Star*, 19 July 1987, B5. This report also noted that "Cominco has been a money-loser four of the past five years." In 1985 its net loss amounted to $97.2 million (figure in "People power could pull the plug on sale of BC utility," *Toronto Star*, 23 January 1987, A20). By the mid-1990s, Cominco was once again operating at a profit; see for example "Cominco posts $152-million profit," *Globe and Mail*, 31 January 1997, B7.

16 The controversy generated considerable publicity in local and national media. The following accounts represent a variety of views, and have informed my own analysis: "Cominco sells electrical utility," *Energy Analects*, Vol. 15 (25 August 1986): 4; "Consumer association opposes sale of West Kootenay to US company," *Energy Analects*, Vol. 15 (8 December 1986): 5; "Utilicorp's plans to purchase West Kootenay receiving

opposition," *Energy Analects*, Vol. 16 (2 February 1987): 6; "BC Utilities Commission okays Cominco's sale of West Kootenay," *Energy Analects*, Vol. 16 (20 July 1987): 5-6; "Cominco sells hydro utility to US firm," *Globe and Mail*, 19 August 1986, B3; "BC utility consumers oppose US takeover," *Globe and Mail*, 29 November 1986, B14; "Cominco sale of unit gets Ottawa okay," *Globe and Mail*, 1 January 1987, B10; "US bid for BC power company stirs hornets' nest of nationalism," *Globe and Mail*, 19 January 1987, A4; "Commission approves sale of BC utility," *Globe and Mail*, 11 July 1987, B2; "People power could pull the plug on sale of BC utility," *Toronto Star*, 23 January 1987, A20; "Selling BC utility to a US company outrages customers," *Toronto Star*, 13 July 1987, B12; "Anger greets US takeover of BC utility," *Toronto Star*, 19 July 1987, B1, B5; "Protest grows over bid to sell power company," *Vancouver Sun*, 20 January 1987, B2; "MLA wants local public ownership of power firm," *Vancouver Sun*, 23 January 1987, A8; John Condit, "Canadian autonomy compromised," *Vancouver Sun*, 26 January 1987, B5.

17 Coverage of CP's sale of Cominco includes Laima Dingwall, "Teck-led group buys control of Cominco from parent CP," *Northern Miner*, Vol. 72, No. 30 (6 October 1986): 1-2; "Teck takeover of Cominco may be record mining coup," *Northern Miner*, Vol. 72, No. 33 (27 October 1986): 1, 20; John Twigg, "Norman Goes Bigtime—Teck Takes Cominco with a Longterm View," *Equity*, Vol. 4, #8 (November 1986): 19-28; William Annett, "The Tecking of Cominco," *BC Business*, Vol. 14, #12 (December 1986): 24-31.

18 These comments are largely drawn from the company history, *Power: The Story of Missouri Public Service*, Kansas City, Missouri: Missouri Public Service, 1992, esp. 121-23.

19 For these conditions, see "Anger greets US takeover of BC utility," *Toronto Star*, 19 July 1987, B1, and "BC Utilities Commission okays Cominco's sale of West Kootenay," *Energy Analects*, Vol. 16 (20 July 1987): 5-6.

20 British Columbia Utilities Commission, "In the Matter of the Utilities Commission Act . . . ," 22.

21 The Conservative government in effect re-regulated other industries, notably the pharmaceutical industry with legislation which targeted generic drug manufacturers.

22 See "Victoria to pull the plug on Hydro's monopoly," *Vancouver Sun*, 18 January 1997, A1, A16. As the article notes, a US ruling—denying BC Hydro a license to buy and sell electricity south of the border—was a key factor in the government's decision (see also "US ruling upends BC Hydro bid," *Globe and Mail*, 18 January 1997, B3). Ironically, just six weeks earlier, the head of BC Hydro voiced his opposition to de-regulation ("Hydro won't allow line use for open market, chair says," *Vancouver Sun*, 6 December 1996, D3).

23 I am grateful to Josh Smienk for providing me with much information about the genesis of the Columbia Basin Trust.

24 "Columbia Basin Trust Act," Chap. 49, *Statutes of British Columbia, 1995*, Victoria: Queen's Printer, 1996. The act came into force on 6 July 1995.

BIBLIOGRAPHICAL ESSAY

This book is based largely on research that I undertook in the archives of West Kootenay Power, and what formerly were its two parent companies, the Consolidated Mining and Smelting Company of Canada (or Cominco) and the Canadian Pacific Railway. The CPR Archives are in Montreal; those of West Kootenay Power are in the company's South Slocan offices (although a few files are also held in the archives of the Rossland Historical Museum and the Nelson Museum); and the Cominco Papers are in the BC Archives and Records Service in Victoria. I have also read widely in contemporary newspapers, notably those of Rossland and Nelson, and explored the relevant government records (especially those of BC's Public Utilities Commission). Interviews with former West Kootenay Power employees proved extremely valuable for giving me a vivid sense of the company's history, and I am grateful that the company had the foresight to record the memories of twenty-six people in the early 1970s. Transcripts of these interviews survive in the Trail office of West Kootenay Power.

I have also been guided by the work of numerous other historians, which has enabled me to think about the history of West Kootenay Power as a regional example of the general process of electrification. The following comments suggest some of my more important debts.

Anyone with a serious interest in the history of electricity ought to consult the work of Thomas P. Hughes. A good place to start is his definitive monograph, *Networks of Power: Electrification in Western Society, 1880–1930* (Baltimore: Johns Hopkins University Press, 1983). More recent American scholarship includes Carolyn Marvin's fascinating book, *When Old Technologies Were New: Thinking about Electric Communication in the late*

Nineteenth Century (Oxford: Oxford University Press, 1988), David E. Nye's *Electrifying America: Social Meanings of a New Technology, 1880–1940* (Cambridge, MA: MIT Press, 1990), and the very readable biography by Neil Baldwin, *Edison: Inventing the Century* (New York: Hyperion, 1995). Two intriguing books on design and society—with considerable detail on electrical objects—are well worth reading: Adrian Forty, *Objects of Desire: Design and Society, 1750–1980* (London: Thames and Hudson, 1986) and Cecelia Tichi, *Shifting Gears: Technology, Literature, Culture in Modernist America* (Chapel Hill: University of North Carolina Press, 1987). The more technically inclined will find much to interest them in W. D. Hackmann, *Electricity from Glass: The History of the Frictional Electrical Machine 1600–1850* (Alphen aan den Rijn, The Netherlands: Sijthoff & Noordhoff, 1978); J. L. Heilbron, *Electricity in the 17th and 18th Centuries: A Study of Early Modern Physics* (Berkeley: University of California Press, 1979); W. James King, *The Development of Electrical Technology in the 19th Century* (Washington DC: US Government Printing Office, 1961, Bulletin 228, Paper 28); Louis C. Hunter and Lynwood Bryant, *A History of Industrial Power in the United States, 1780–1930: Vol. 3, The Transmission of Power* (Cambridge, MA: MIT Press, 1991); and the two articles by C. Mackechnie Jarvis, "The Generation of Electricity" and "The Distribution and Utilization of Electricity," both of which appear in Charles Singer, E. J. Holmyard, A. R. Hall, and Trevor I. Williams, editors, *A History of Technology*, Vol. V, The Late Nineteenth Century, *c.* 1850–*c.* 1900 (Oxford: Clarendon Press, 1958), 177-234. In addition, Arnold Roos has compiled two excellent bibliographies that list much of the literature on hydro-electric engineering in Canada: see *Select Bibliography of Engineering Literature Concerning Electrical and Hydraulic Engineering, Electrical Transmission and Canadian Hydro-Electric Installations, 1884–1955* (Ottawa: Environment Canada—Parks Service, 1992, Microfiche Report Series 458) and *A Bibliography of the History of Canadian Science and Technology/Une bibliographie de l'histoire de la science et de la technologie au Canada* (Ottawa: Canadian Science and Technology Historical Association/Association pour l'histoire de la science et de la technologie au Canada, 1995).

The best Canadian work on the history of electricity has come from the pen of H. V. Nelles. *The Politics of Development: Forests, Mines and Hydro-electric Power in Ontario, 1849–1941* (Toronto: Macmillan of Canada, 1974) remains a key text, while a more recent book (co-written with Christopher Armstrong) is also essential reading: *Monopoly's Moment: The Organization and Regulation of Canadian Utilities, 1830–1930* (Philadelphia: Temple University Press, 1986). Another book by Armstrong and Nelles, *Southern Exposure: Canadian Promoters in Latin America and the Caribbean 1896–1930* (Toronto: University of Toronto Press, 1988), provides a fascinating description of Canadian involvement in electrical facilities abroad. Duncan McDowall's *The Light: Brazilian Traction, Light and Power Company Limited, 1899–1945* (Toronto: University of Toronto Press, 1988) is also useful for the light it sheds on this topic. Several of West Kootenay Power's early presidents — William Doull and C. R. Hosmer — invested in Latin American utilities, a good indication of how general such activity was in the early twentieth century. For detail of another West Kootenay president — perhaps the most important figure in the company's pioneer years, the baffling Sir Charles Ross — see the first chapter of Roger Phillips' *A Historical Appraisal of the Ross Rifle* (Antigonish, NS: Casket Printing and Publishing Co., Ltd., 1984). Several essays are valuable for their discussions of the drive for municipal control and ownership of public utilities: see Paul Rutherford, "Tomorrow's Metropolis: The Urban Reform Movement in Canada, 1880–1920," and John C. Weaver, "'Tomorrow's Metropolis' Revisited: A Critical Assessment of Urban Reform in Canada, 1890–1920," both in Gilbert A. Stelter and Alan F. J. Artibise, eds., *The Canadian City: Essays in Urban and Social History* (Ottawa: Carleton University Press, 1984).

British Columbian historians have written little about energy, and the only general overview is now somewhat dated: Mary Doreen Taylor, "Development of the Electricity Industry in British Columbia" (MA thesis, University of British Columbia, 1965). Patricia Roy's dissertation, "The British Columbia Electric Railway Company, 1897–1928: A British Company in British Columbia" (PhD thesis, University of British Columbia, 1970), is a valuable introduction to the Lower Mainland's dominant private utility. She produced a number of useful studies based on this unpublished work: "Regulating the British Columbia Electric Railway: The First Public Utilities Commission in British Columbia" (*BC Studies*, #11 [Fall 1971]: 3-20); "The Fine Arts of Lobbying and Persuading: The Case of the British Columbia Electric Railway, 1897–1917," in David S. Macmillan, ed., *Canadian Business History: Selected Studies 1497–1971* (Toronto: McClelland and Stewart, 1972, 239-54); "The British Columbia Electric Railway and Its Street Railway Employees: Paternalism in Labour Relations" (*BC Studies*, #16 [Winter 1972–73]: 3-24); and "Direct Management from Abroad: The Formative Years of the British Columbia Electric Railway" (*Business History Review*, Vol. 47, No. 2 [Summer 1973]: 239-59). Professor Roy examines an earlier period in her article, "The Illumination of Victoria: Late Nineteenth-Century Technology and Municipal Enterprise" (*BC Studies*, #32 [Winter 1976–77): 79-92]. A similar study by G. W. S. Brooks is also noteworthy: "Edgar Crow Baker: An Entrepreneur in Early British Columbia" (*BC Studies*, #31 [Autumn, 1976]: 23-43). Much information on this era — and some intriguing photographs — may be found in Douglas Parker's books, *No Horsecars In Paradise: A History of the Street Railways and Public Utilities in*

Victoria, British Columbia before 1897 (Toronto: Railfare Enterprises, 1981) and *Streetcars in the Kootenays: Nelson's Tramways, 1899–1992* (Edmonton: Havelock House, 1992). Also useful for understanding events in Nelson is Patrick Wolfe's study of the redoubtable Houston: "Tramp Printer Extraordinary: British Columbia's John "Truth" Houston" (*BC Studies*, #40 [Winter 1978–79]: 5-31), although this should be read with Rutherford's and Weaver's work, cited above.

Rather more has been written about the province's recent past, largely because of the controversial nature of the relationship between politics and energy. For the Bennett years and the formation of BC Hydro, see Paddy Sherman, *Bennett* (Toronto: McClelland and Stewart, 1966); Martin Robin, *Pillars of Profit: The Company Province, 1934–1972* (Toronto: McClelland and Stewart, 1973); and H. William Tieleman, "The political economy of nationalization: Social Credit and the takeover of the British Columbia Electric Company" (MA thesis, University of British Columbia, 1984). Another key aspect of the province's post-war energy policy is explored in John R. Wedley's "Infrastructure and Resources: Governments and their Promotion of Northern Development in British Columbia, 1945–1975" (PhD thesis, University of Western Ontario, 1986). David J. Mitchell's biography of Bennett, *W. A. C. Bennett and the Rise of British Columbia* (Vancouver: Douglas & McIntyre, 1983), provides the former premier a platform from which to defend and justify his various political manoeuvres, and should only be read in conjunction with other, more critical accounts. The Columbia River Treaty of 1961 generated a good deal of controversy as well as much written commentary. Neil A. Swainson provides a good overview in *Conflict over the Columbia: The Canadian Background to an Historic Treaty* (Montreal: McGill-Queen's University Press, 1979) as well as a comprehensive bibliography. His approach tends to discount the criticisms of the treaty by people such as Larratt Higgins, whose several articles offer a very different analysis. See especially "The Columbia River Treaty: A Critical View," *International Journal* (Vol. 16, No. 4 [Autumn 1961]: 399-404), and "The Alienation of Canadian Resources; The Case of the Columbia River Treaty," in Ian Lumsden, ed., *Close the 49th Parallel Etc.* (Toronto: University of Toronto Press, 1973, 223-40). A third perspective — that of local residents who objected to the treaty's provision of the High Arrow dam in particular, with its subsequent flooding of the Arrow Valley — may be found in Donald Cresswell Waterfield, *Continental Waterboy: The Columbia River Controversy* (Toronto: Clarke, Irwin, 1970); J. D. McDonald, *Storm over High Arrow: The Columbia River Treaty (A History)* (Rossland: privately printed, 1993); and Richie Deane's submission to the External Affairs Committee of the House of Commons, 30 March 1964, "The Columbia Treaty and High Arrow Dam." Another valuable source for those interested in understanding the complex negotiations surrounding the treaty is the series of public lectures entitled "Canadian-American Relations: Perspectives on the Columbia River Treaty," delivered in the spring of 1974 at Simon Fraser University. The lectures, fourteen in all, include contributions by many of the participants in the treaty negotiations, and survive on audiocassette tapes in the Simon Fraser University Archives.

More generally, the footnotes accompanying the text suggest my reliance on a wide range of primary and secondary accounts. I should add that this book also reflects my earlier research on the history of the Kootenays, published as *Roaring Days: Rossland's Mines and the History of British Columbia* (Vancouver: UBC Press, 1995). An extensive bibliography was included in that volume, which may be of interest to those who wish to study the area and its historiography in greater detail.

APPENDICES

West Kootenay Power Company Ltd. Transmission Lines Historical Data

Org. Line No.	Pres. Line No.	Year Compl.	Volts. (KV)	From	To	Line Lgth. (km)
1		1898	22	Old No. 1 Plant (Lower Bonnington Falls)	Rossland	51.48
		1906		Dismantled		
2		1899	50	Old No. 1 Plant (Lower Bonnington Falls)	Castlegar–Tadanac (Trail)–Rossland	50.2
		1926		Tadanac to Rossland section removed		
		Jan. 1937		Line disconnected at Tadanac. Bonnington to Castlegar section used to feed Castlegar Substation		
		Sept. 1937		Remaining line disconnected and copper wire removed		
Trail Branch Line		1899	20	On mainline 2.91 miles from Rossland	Old Trail Substation	4.99
		1928		Line removed		
3 Double-circuited with original #2		1906	20	Old No. 1 Plant (Lower Bonnington Falls)	Castlegar–Tadanac (Trail)–Rossland	50.2
		1926		Tadanac to Rossland section removed		
		Jan. 1937		Line disconnected at Tadanac. Bonnington to Castlegar section used to feed Castlegar Substation		
		Sept. 1937		Remaining line disconnected and copper wire removed		
Second No. 1		1906	63	No. 2 Plant (Upper Bonnington Falls)	Anderson's Ranch (Trail)–Rossland	51.49
				Rossland	Grand Forks	51.97
				Grand Forks	Phoenix	14.48
				Phoenix	Greenwood	7.08
						125.02
		1920		Added: Greenwood	Oliver	73.08
				Oliver	Allenby	88.85

		1928		Removed Anderson's Ranch	Rossland	
		1939		Added new section: Tadanac	Rossland	8.51
	2	1905		Re-numbered: South Slocan	Tadanac	44.59
	10	1919		Tadanac	Allenby–Copper Mountain	235.46
		1948		Re-terminated at Warfield Switching Station	Oliver	
4		1906	63	No. 2 Plant (Upper Bonnington Falls)	Anderson's Ranch (Trail)–Rossland	51.49
				Rossland	Grand Forks	51.97
				Grand Forks	Phoenix	14.48
				Phoenix	Greenwood	7.08
						125.02
		1920		Added section: Greenwood	Oliver	73.08
		1927		Restrung: South Slocan	Tadanac	
		1928		Removed Anderson's Ranch	Rossland	
				New section: Tadanac	Rossland	8.51
					Tadanac	
	2	1939		Re-numbered: South Slocan	Tadanac	152.4
	9	1919		Tadanac	Oliver	
		1948		Re-terminated at Warfield Switching Station	Oliver	
5		1925	63	Upper Bonnington Plant	#2 Sub. (Trail)	44.59
		1928		Re-terminated at South Slocan	Tadanac Switching Station	
	1	1939		Line was re-numbered		
6		1931	63	South Slocan Plant	Tadanac Switching Station	44.59
	5	1939		Line was renamed		
7		1937	63	Replaced original #2 & #3 lines and started at South Slocan	Tadanac Switching Station	44.59

West Kootenay Power Company Ltd. Transmission Lines Historical Data *continued*

Org. Line No.	Pres. Line No.	Year Compl.	Volts. (KV)	From	To	Line Lgth. (km)
	3	1939		Line was re-numbered		
	6	1940	63	South Slocan (#3) Plant	Tadanac Switching Station	44.59
	7	1944	63	#5 Plant (Brilliant	Tadanac Switching Station	28.48
	8	1949	63	#5 Plant (Brilliant	Tadanac Switching Station	28.48
	11	1963/64	170	Warfield Switching Station	Oliver	138.78
	13	1974	63	South Slocan (#3) Plant	BCH Canal Plant	0.40
	14	1966	63	Warfield Switching Station	Waneta (#6) Plant	17.22
	15	1953	63	Warfield Switching Station	Waneta (#6) Plant	17.22
	16	1953	63	Warfield Switching Station	Waneta (#6) Plant	17.22
	17	1953	63	Tadanac Switching Station	Waneta (#6) Plant	17.7
	18	1975	63	Waneta (#6) Plant	Seven Miles Dam site	9.33
		1985 upgrade		Waneta (#6) Plant	#20 Line	7.93
	19	1958	63	South Slocan (#3) Plant	Slocan City	40.23
	20	1931	63	Tadanac Switching Station Fruitvale	Fruitvale Salmo	24.14 22.21
	21	1924	63	Lower Bonnington (#1) Plant	South Slocan (#3) Plant	1.77
	22		63	Upper Bonnington (#2) Plant	South Slocan (#3) Plant	3.37
	23		63	Corra Linn (#4) Plant	Upper Bonnington Plant	1.66
				Upper Bonnington (#2) Plant	Lower Bonnington Plant	1.6
				Lower Bonnington (#1) Plant	South Slocan (#3) Plant	1.77
	24		63	South Slocan (#3) Plant	Corra Linn (#4) Plant	5.03
	25		63	South Slocan (#3) Plant	Brilliant (#5) Plant	16.41
	26	1960	63	Brilliant (#5) Plant	Saw Mill in Castlegar	10.22

	27	1928	63	Salmo	Corra Linn (#4) Plant	52.79
	28	1938	63	Upper Bonnington (#2) Plant	City of Nelson's Bonnington Switchyard	5.15
	29	1957	138	South Slocan (#3) Plant	BC Hydro Whatshan Plant	39.4 WKP section
	30W	1952	170	South Slocan (#3) Plant	Coffee Creek Sub/West Tower	58.74
	Crossing	Apr. 8, 1952	170	Kootenay Lake Crossing		3.54
	30E	1952	170	East Tower–Crawford Bay	Kimberley	75.8
	31	1953	63	A. A. Lambert Terminal, Creston	Creston Sub	6.69
	32	1953	63	Crawford Bay	A. A. Lambert Terminal	73.00
	37	1963	63	Coffee Creek Sub	Kaslo	23.54
38	out	1962	63	Coffee Creek: Lake Crossing	Crawford Bay	3.54
	40	1964	161	Oliver Terminal	R. G. Anderson Terminal, Penticton	35.45
1	41	1921	63	Oliver Terminal	Waterford Sub	Total
				Waterford Sub	R. G. Anderson Terminal, Penticton	35.34
2	42	1921	63	Oliver Terminal	Huth Avenue Sub	36.24
43		1936	63	Oliver Terminal	Princeton	117.80
	43	1994 upgrade	138	Oliver Terminal	Princeton	117.80
	43A	1986	138	Hedley Sub	Mascot Mine	4.99
	44	1936	63	Oliver Terminal–Pine Street Sub	Osoyoo	22.21
	45	1922	63	R. G. Anderson Terminal, Penticton	Westminster Sub	4.62
				Westminster Sub		
				Naramata	Naramata	11.27
					D. G. Bell Terminal	31.70

West Kootenay Power Company Ltd. Transmission Lines Historical Data *continued*

Org. Line No.	Pres. Line No.	Year Compl.	Volts. (KV)	From	To	Line Lgth. (km)
	46	1958	138	F. A. Lee Terminal, Kelowna	BC Hydro, Vernon	43.13
	47	1965		R. G. Anderson Terminal, Penticton	D. G. Bell Terminal	
	48	1936	63	F. A. Lee Terminal, Kelowna	BC Hydro, Vernon	46.67
	49	1921	63	Huth Avenue Sub, Penticton	Summerland	16.06
50		1922	63	F. A. Lee Terminal, Kelowna	OK Mission–D. G. Bell	
	50	upgraded	138	F. A. Lee Terminal, Kelowna	OK Mission, Kelowna	18.59
51		1959	138	F. A. Lee Terminal, Kelowna	D. G. Bell Terminal	
	51	upgraded	138	F. A. Lee Terminal, Kelowna	D. G. Bell–OK Mission	25.66
	52	1966	63	Huth Avenue Sub, Penticton	R. G. Anderson Terminal	3.86
	53	1966	63	Huth Avenue Sub, Penticton	R. G. Anderson Terminal	3.86
55		1976	63	F. A. Lee Terminal, Kelowna	Hollywood–Harvey Avenue	
	55	upgraded	138	F. A. Lee Terminal, Kelowna	Springfield Road	14.00
	56	1980	138	Princeton	BCH Line to Smilkameen	6.13
	60		63	Oliver Terminal	Osoyoos	
	72	1975	230	F. A. Lee Terminal, Kelowna	BC Hydro, Vernon	43.13
	73	1985	230	R. G. Anderson Terminal, Penticton	F. A. Lee Terminal	58.26
	74	1996	230	F. A. Lee Terminal, Kelowna	BC Hydro, Vernon	

Source: West Kootenay Power Archive—Historic Data Binders
"Documentation of Line Numbers": 04-09-1989; J. M. Detmold
"WKP/Cominco Transmission Line Information": 7-16-1996; W. Zhang

West Kootenay Power Company Ltd. Generating Plants and Equipment Data

Plant	Unit	Head (Feet)	Year Installed	Generator	Turbine
Original #1 Bonnington Falls	1	34	1897	CGE, 750 kW, 1100 volts, 3 phase, 60 cycles, 180 RPM, horizontal direct connected	Stillwell-Bierce & Smith-Vaile, Victor Type, horizontal, twin runners, 39″ diameter wheels, 1184 HP
	2	34	1897	CGE, 750 kW, 1100 volts, 3 phase, 60 cycles, 180 RPM, horizontal direct connected	Stillwell-Bierce & Smith-Vaile, Victor Type, horizontal, twin runners, 39″ diameter wheels, 1184 HP
	3	34	1899	CGE, 1500 kW, 1100 volts, 3 phase, 60 cycles, 180 RPM, horizontal direct connected	Stillwell-Bierce & Smith-Vaile, Victor Type, horizontal, twin runners, 43″ diameter wheels, 1648 HP
New #1 Lower Bonnington	1	70	Jun. 1925	CGE, 17500 kVA, 7200 volts, 3 phase, 60 cycles, 90% Power Factor, 100 RPM, vertical direct connected	Canadian Allis Chalmers, 20000 HP, Francis type single 120″ diameter runner
	2	70	Aug. 1925	CGE, 17500 kVA, 7200 volts, 3 phase, 60 cycles, 90% Power Factor, 100 RPM, vertical direct connected	Canadian Allis Chalmers, 20000 HP, Francis type single 120″ diameter runner
	3	70	Aug. 1926	CGE, 17500 kVA, 7200 volts, 3 phase, 60 cycles, 90% Power Factor, 100 RPM, vertical direct connected	Canadian Allis Chalmers, 20000 HP, Francis type single 120″ diameter runner
Original #2 Upper Bonnington	1	70	Nov. 1916	CGE, Type ATB, 6700 kVA, 2300 volts, 3 phase, 60 cycles, 180 RPM, vertical direct connected Re-wound & re-piled in 1928 to increase output to 7500 kVA	Canadian Allis Chalmers, 9000 HP, Francis type vertical, triple runners, 65″ diameter, bronze
	2	70	1907	CGE, Type ATB, 6000 kVA, 2300 volts, 3 phase, 60 cycles, 40 poles, 180 RPM, vertical direct connected Re-wound & re-piled in 1928 to increase output to 7000 kVA	I. P. Morris, 8000 HP, Francis Type, vertical, triple runner, 60″ diameter wheels, bronze

West Kootenay Power Company Ltd. Generating Plants and Equipment Data *continued*

Plant	Unit	Head (Feet)	Year Installed	Generator	Turbine
	3	70	1907	CGE, Type ATB, 6000 kVA, 2300 volts, 3 phase, 60 cycles, 40 poles, 180 RPM, vertical direct connected Re-wound & re-piled in 1928 to increase output to 7000 kVA	I. P. Morris, 8000 HP, Francis Type, vertical, triple runner, 60″ diameter wheels, bronze
	4	70	1916	CGE, Type ATB, 6700 kVA, 2300 volts, 3 phase, 60 cycles, 40 poles, 180 RPM, vertical direct connected Re-wound & re-piled in 1928 to increase output to 7500 kVA	Canadian Allis Chalmers, 9000 HP, Francis type vertical, triple runners, 65″ diameter, bronze
#2 Upper Bonnington Extension	5	70	Mar. 1940	CWC, 17500 kVA, vertical direct connected, 7200 volts, 90% power factor, 3 phase, 60 cycles, 100 RPM, continuous overload capacity = 20500 kVA	Canadian Allis Chalmers, 26000 HP, Francis type, vertical, single runner. The turbine is capable of generating 28500 HP @ 100 RPM and 75 feet head.
	6	70	Mar. 1940	CWC, 17500 kVA, vertical direct connected, 7200 volts, 90% power factor, 3 phase, 60 cycles, 100 RPM, continuous overload capacity = 20500 kVA	Canadian Allis Chalmers, 26000 HP, Francis type, vertical, single runner. The turbine is capable of generating 28500 HP @ 100 RPM and 75 feet head.
#3 South Slocan	1	70	Dec. 1928	CGE, 17500 kVA, 7200 volts, 3 phase, 60 cycles, 90% power factor, 100 RPM, vertical connected	Canadian Allis Chalmers, 25000 HP, Francis type, vertical, single runner, 128″ throat diameter
	2	70	Dec. 1928	CGE, 17500 kVA, 7200 volts, 3 phase, 60 cycles, 90% power factor, 100 RPM, vertical connected	Canadian Allis Chalmers, 25000 HP, Francis type, vertical, single runner, 128″ throat diameter
	3	70	Jun. 1929	CGE, 17500 kVA, 7200 volts, 3 phase, 60 cycles, 90% power factor, 100 RPM, vertical connected	Canadian Allis Chalmers, 25000 HP, Francis type, vertical, single runner, 128″ throat diameter
#4 Corra Linn	1	53	1932	CGE, 15000 kVA, 7200 volts, 3 phase, 60 cycles, 90% power factor, 85.7 RPM, vertical connected	Dominion Engineering Works, 19000 HP, Francis type, I. P. Morris, vertical single runner, 123.25″ throat diameter

	2	53	1932	CGE, 15000 kVA, 7200 volts, 3 phase, 60 cycles, 90% power factor, 85.7 RPM, vertical connected	Dominion Engineering Works, 19000 HP, Francis type, I. P. Morris, vertical single runner, 123.25″ throat diameter
	3	53	1932	CGE, 15000 kVA, 7200 volts, 3 phase, 60 cycles, 90% power factor, 85.7 RPM, vertical connected	Dominion Engineering Works, 19000 HP, Francis type, I. P. Morris, vertical single runner, 123.25″ throat diameter

West Kootenay Power Company Ltd. Early Day Substations, Switching Stations Designed and Constructed by WKP&L

Stn.	Loc./ Supply	Date in Op.	Cap. (kVA)	Transformation and Associated Equipment
Original Rossland Substation	Monte Christo Mine/Rossland & mines	1897	3750	2 banks of three CGE, single phase, 242 kW, air blast, 16000/2200 volts, Y-Delta connected transformers
Second Rossland Substation	Idaho Mineral Claim/Rossland & mines	April 1898		Equipment moved from the original Rossland Substation
		1906		Two CWC single phase, 1250 kW, 60000/2200 volts, OIWC transformers installed
		1909		Third CWC single phase, 1250 kW, 60000/2200 volts, OIWC transformer added
		1929		Substation was dismantled
Nelson Substation	Nelson/Nelson Lighting & Tramway and Hall Mines Smelter	1899	1452	2 banks of three CGE, single phase, 242 kW, air blast, 16000/2200 volts, Y-Delta connected transformers Station was later taken out of service, then was demolished by fire
Grand Forks Substation	Grand Forks	1906	1500	1 bank of three CGE, single phase, 242 kW, 20000/500 volts, air blast, Delta-Delta connected transformers 1 bank of three CWC, single phase, 242 kW, 20000/500 volts, OISC, Delta-Delta connected transformers
		1929		Added 1 bank of 3 CGE, single phase, outdoor type, 500 kVA, type H, 60000/2200 volts with 2300 volts tertiary, OISC, Delta-Delta connected transformers
Greenwood Substation	Greenwood/Area mines & smelter	1906	3750	1 bank of three CWC, single phase, indoor type, 1250 kVA, 60000/2200 volts, OIWC transformers 1 bank of three CWC, single phase, 312 kW, 20000/2200 volts, Y-Y connected transformers supplying BC Copper Co.—later the bank was reconnected to 2300/10000 volts, Y-Deltya connected to supply Dentonia Mine

Phoenix Substation	Phoenix/Grandy Mines and others	1936	3750	1 bank of three CWC, single phase, indoor type, 1250 kW, 60000/2200 volts, OIWC, Delta-Delta connected transformers Standby Equipment: 3 banks of eight indoor type, WMC single phase, 312 kW, 20000/2200 volts, OISC transformers Station is now dismantled
Silver King Mine Substation	Nelson/ Silver King Mine	1908	726	550 volts transformers taken from Canada Zinc Co. substation
#1 Substation, Tadanac	CM&S Smelter (Cominco)	1916	22500	3 banks of three CWC, single phase, 2500 kVA, 60000/2200 volts, OIWC, Delta-Delta connected transformers
Northport Substation	Northport, Washington, USA	Nov. 1918 Dec. 1942 1949	4500	1 bank of three CGE, single phase, 1500 kVA, OIWC, indoor type, 60000/2200 volts, Delta-Delta connected transformers Station de-energized Station was dismantled
Copper Mountain Substation	Near Princeton/Mines	1919 1929 1939	2250	1 bank of three CWC, single phase, 750 kVA, 60000/2200 volts, OISC, outdoor type, Delta-Delta connected transformers CGE Synchronous condenser installed Station dismantled and equipment moved to Oliver substation
Allenby Substation	Near Princeton/ Mines	1920	7500	2 banks of three CWC, single phase, 1250 kVA, 60000/2200 volts, OIWC, Delta-Delta connected transformers
Oliver Outdoor Substation	Oliver/ Oliver and district	1922 1937 1939	1800	1 bank of three CGE, single phase, 600 kVA, 60000/22000/2200 volts, H-60, outdoor type, Delta-Delta connected transformers Outdoor Switching Station Added Rebuilt 2000 kVA synchronous condenser moved from Allenby Substation with three 750 kVA single phase, 60000/2300 volts, OISC transformers
Kelowna Substation	Mission District in Kelowna	1922	1875	6 single phase, 312.5 kW, indoor self-cooled transformers supplying power to the city

West Kootenay Power Company Ltd. Early Day Substations, Switching Stations Designed and Constructed by WKP&L *continued*

Stn.	Loc./ Supply	Date in Op.	Cap. (kVA)	Transformation and Associated Equipment
Kelowna Substation (continued)		1932	600	Station was revampled with three CWC 200 kVA, 2300/7200 volts, outdoor type OISC transformers installed for country circuit
		1936	1500	Three Moloney Electric Co. single phase, 500 kVA, 33000/2300 volts, OISC transformers bought from BC Power Commission, used as tied in with West Canadian Hydro
		1937	4500	New steel substation erected and voltage raised. Three Packard Electric Co., single phase, 1500 kVA, 6000/2300 volts, outdoor type, OISC transformers installed
		1942		33000/2300 volts Moloney Electric Co. transformers removed
		1947		Second bank of three single phase, 200 kVA, 2300/7200 volts, transformers added for Kelowna country circuit
		1948		A 200 kVA booster transformer placed in city circuit to boost voltage
		1949		Added Synchronous Condenser moved from Allenby Substation
Penticton Substation	Penticton/ Municipality of Penticton, Kelowna Mission Substation and Summerland line	1922	1500	One bank of three single phase, 500 kVA, 6000/5000 volts, transformers installed
				One bank of three single phase, 500 kVA, 5000/20000 volts, transformers installed to feed Kelowna Mission Substation
		1933		Above bank used to feed Summerland Line
		1948		New substation in operation. One bank of three Packard Electric Ltd., single phase, 1500 kVA, 6000/4600/2300 volts, 40° rise, OISC, outdoor type transformers
#2 Substation (was #3)	CM&S Lead Refinery, Tadanac (Cominco)	1926	15000	1 bank of three CGE single phase, 5000 kVA, 6000/2200 volts, outdoor type, oil cooled, Delta-Delta connected transformers

#3 Substation (was #2)	CM&S Smelter/ #3 Rotary Rm., Old CM&S Zinc Plant	1929	51500	10 CGE, three phase, 2600 kVA, 6000/380 volts, transformers, Delta-six phase star connection 1 bank of three Brown-Boveri single phase, 8500 kVA, Delta-Delta connected transformers. For mercury arc rectifiers
Outdoor Rossland Substation	Old Rossland Club site/Rossland & district	1930 1949	1500	1 bank of three Maloney single phase, 500 kVA, 6000/2200 volts, OISC, Delta-Delta connected transformers with one WMC, 60 kVA, 3 phase, 10% regulation regulator Second regulator installed: CWC, 40 kVA, 3 phase, 10% regulation
#4 Substation, Tadanac	CM&S Smelter/Part of Smelter and Slag Retreatment Plant	Jan. 1931	15000	1 bank of three CGE single phase, 5000 kVA, 6000/2300 volts, outdoor type, self-cooled/water cooled, Delta-Delta connected transformers
#5 Substation, Warfield	CM&S Warfield/ Fertilizer Plant	Jan. 1931 1932 1941	15000 17500 30000	1 bank of three CGE single phase, 5000 kVA, 6000/2300 volts, OISC/OIWC, outdoor type, Delta-Delta connected transformers One CGE, 5000 kVA transformers added. Together with the previously installed three single phase transformers formed tow open Delta connected banks Two additional CGE single phase, 5000 kVA, transformers added. Together with previously installed transformers formed two banks of three single phase transformers, Delta-Delta connected
#6 Substation, Warfield	CM&S Warfield/ hydrogen rectifier units	1931	21000	1 three phase, CGE, 7000 kVA, 6000/1040 volts, OISC/WC transformer 2 three phase, Brown-Boveri, 6000 kVA, 6000/1040, OISC/WC transformers
Kaleden Substation	Kaleden/Kaleden and areas	1932	200	2 CGE, single phase, 100 kVA, 6000/2300 volts, OISC, outdoor type transformers
Summerland Substation	Summerland	1933 1949	600 1500	1 bank of three Packard Electric single phase, 200 kVA, 22000/5000 volts, OISC transformers New station: 1 bank of three single phase, 500 kVA, 60000/5000 volts, OISC, outdoor type transformers using Packard cores in Westinghouse tanks
Kootenay Belle Substation	Sheep Creek/area mines and mills	1934 later	3000 600	Three 1000 kVA transformers, later moved to Castlegar Substation 1 CGE three phase, 600 kVA, 6000/2300 volts, OISC, outdoor type transformer

West Kootenay Power Company Ltd. Early Day Substations, Switching Stations Designed and Constructed by WKP&L *continued*

Stn.	Loc./ Supply	Date in Op.	Cap. (kVA)	Transformation and Associated Equipment
Keremos Substation	Keremeos	1934	300	3 CGE single phase, 100 kVA, 6000/2300 volts, OISC, 40°C rise, outdoor type transformers
Salmo	Salmo	1934	300	Original installation: One 300 kVA, 3 phase transformer
		1934	100	300 kVA transformer was replaced with one 100 kVA, single phase transformer, 6000/2300 volts
		1952	300	Transformer was replaced with one 300 kVA transformer
		1954	600	Transformer was replaced with one 600 kVA unit
Ymir Substation	Ymir	1934	600	Two 300 kVA, three phase transformers
		later	100	Original transformers were replaced with one 100 kVA, 60000/2300 volts, OISC single phase transformer
#7 Substation, Stoney Creek	Stoney Creek, Trail/Stoney Creek Pump Station	1935	9000	1 bank of three Packard Electric single phase, 3000 kVA, 60000/2200 volts, OISC/WC transformers
#8 Substation, Tadanac	CM&S Smelter/ Zinc Plant	1935	15000	1 bank of three CGE single phase, 5000 kVA, 60000/2200 volts, OISC/WC transformers
Hedley	Hedley/Hedley Mascot Mine & Town of Hedley	1935	3750	1 bank of three CWC single phase, 1250 kVA, 60000/2200/440 volts, OIWC transformers. 1 spare transformer with same rating as above
Greenwood Outdoor Switching Station	Greenwood	1936		

Naramata Substation	Naramata, Okanagan	1937	300	Three single phase, 100 kVA, 60000/2300 volts, OISC, outdoor type transformers
Rotter's Mill Substation	Near Salmo/ F. R. Rotter's Sawmill	1940	300	1 CGE, three phase, 300 kVA, 60000/2300 volts, OISC transformer
Emerald Mine Substation	North of Salmo/ Emerald Mine	Mar. 1943	1200	Two 300 kVA single transformers and one 600 kVA, three phase transformer
		Sept. 1943		Mine and mill shut down, station de-energized
		later	4500	Mine re-opened by Canada Exploration, station re-energized with bank of three General Electric single phase, 1500 kVA, 60000/2300 volts, OIWC, indoor type transformers
Midway	Midway	1946	100	1 CGE single phase, 100 kVA, 60000/2300 volts, OISC transformer
Okanagan Falls Substation	Okanagan Falls/ areas	1946	200	2 CGE single phase, 100 kVA, 60000/2300 volts, OISC, outdoor type transformers
Kenville Substation	Taghum near Nelson/Kenville Mine and Mill	1947	900	2 CGE single phase, 300 kVA, 60000/2300 volts, transformers plus 1 CGE single phase, 300 kVA, 60000/480 volts, transformer
Princeton Substation	Princeton/ Princeton and area	1947	600	1 Packard Electric three phase, 600 kVA, 60000/2300/460 volts, OISC, outdoor type transformer
Reeves MacDonald Mine Substation	Near Salmo/Reeves MacDonald Mine	1948	2400	2 General Electric single phase, 1200 kVA, 60000/2300 volts, OISC, outdoor type transformers
H. B. Mine Substation	Salmo/H. B. Mine	1949	600	1 Packard Electric three phase, 600 kVA, 6000/2300 volts, OISC, outdoor type transformer
Rutland Substation	Kelowna/ Rutland areas	1949	1500	1 single phase transformer with Packard Electric tank and core, 500 kVA, 6000/5000 volts, OISC outdoor type. 2 single phase transformers with Packard Electric tanks and Westinghouse cores, 500 kVA, 6000/5000 volts, OISC outdoor type. The transformers were connected in Delta-Y

Legend: CGE = Canadian General Electric Co. CWC = Canadian Westinghouse Co. WMC = Westinghouse Manufacturing Co. OIWC = oil immersed, water cooled OISC = oil immersed, self-cooled

INDEX